REVIEW FOR THE SCIENCE SECTION OF THE GED TEST

By
James R. A. Frendak

This book is correlated to the video tapes produced by COMEX Systems, Inc., Review for The GED Science test by Kate Prybylowski ©2001 it may be obtained from

comex systems, inc.

5 Cold Hill Rd.
Suite 24
Mendham, NJ 07945

Published by

comex systems, inc.
5 Cold Hill Rd., Suite 24
Mendham, NJ 07945

ISBN 1-56030-140-6

INTRODUCTION

The purpose of this study guide is to help you prepare for the section of the General Educational Development (GED) Test in Science. This text will not only increase your knowledge of the science related subjects, but will also increase your vocabulary of science words, sharpen your skill at answering science questions, and build your confidence level in the science area.

WHAT SUBJECTS ARE TESTED IN THE SCIENCE PART OF THE GED?

Questions in the science portion of the GED will deal with the life sciences (biology) and the physical sciences (earth science, physics, and chemistry). While it is not important for you to study these subject areas in depth, it is important for you to have a basic knowledge of the important concepts of each one. That is what you will accomplish by using this study material. You will find an overview of the important concepts of each of the subjects as well as a vocabulary list of science words you will find helpful. Many times knowing what a word means will enable you to answer a question. By using this study guide you will learn a great deal and have a better understanding of the material on which you will be tested.

WHAT KIND OF QUESTIONS WILL I HAVE TO ANSWER?

The questions on this test will be the multiple-choice type. You will be given a list of five choices and be asked to choose the best answer to the question. There are a total of fifty questions and you will have ninety-five minutes to complete this part.

All of the questions will relate to one of the science areas. You will be given a short passage, diagram, or chart and be asked a question or questions that relate to what you read or can infer from the information. For this reason, the **MOST IMPORTANT SKILL** needed for this test is your skill of **READING COMPREHENSION**. If you are unable to read and understand the question, you will not be able to answer the question. However, if you are able to understand the question, in many cases you will be able to answer the question even if you had no prior knowledge of the topic. It is therefore important to have a working science vocabulary. We have designed this book to help you develop your vocabulary and to familiarize yourself with science content. We have broken the book into many short lessons and questions. While the passages tend to be a little longer than what you will encounter on the GED, if you can comprehend them, you will do well on the GED. At the end of each main chapter there is a section with questions that are similar to what you will find on the GED as well as a complete test at the back of the book.

PROCEDURE FOR USING THIS STUDY GUIDE

You are now ready to begin your study. Work carefully throughout the book. Become familiar with the vocabulary words. Answer the chapter questions as you read the information and take the chapter tests as you complete each section. Review the areas that you find difficult to understand. Do not concentrate on memorizing information. Remember the science test is based more on your comprehension of the question than it is on material that you have memorized. After you have completed the study material, take the sample science test. As in all of the others sections of the GED examination, an <u>unanswered</u> question is considered a <u>wrong</u> answer. Develop the attitude that you <u>must</u> answer <u>every</u> question. By having such an attitude, you will increase your chances of performing well on this test.

GED – SCIENCE INTRODUCTION

There is no possible way to review all of the science topics in this book. What follows is an overview of what you may need to know in order to prepare for the science questions presented on the GED Science Test. Remember, no one is expected to know everything about science. Don't try and memorize all the material because our only aim is to present a review of science concepts. After reading this book and taking the practice test, you will have an idea of what the GED Science Test is about. You will gain the confidence to attempt the test and achieve a passing score. You have nothing to lose and everything to gain by taking the test. Good luck!

Breakdown of the GED Science Test

GED Test Question Breakdown	Fundamental Understandings	Unifying Concepts & Processes	Science as Inquiry	Science & Technology	Science in Personal & Social Perspective	History & Nature of Science	Total
Life Science	14	1	2	1	4	1	23
Earth & Space	6	0	1	1	2	1	11
Physics & Chemistry	10	1	1	0	2	2	16
Totals	30	2	4	2	8	4	50

Looking at the above, being able to understand basic concepts as they are presented is **very** important. It accounts for thirty questions or 60% of the test. Being able to take science topics and relating them to how they affect you or society is the next most important skill accounting for eight questions, or 16%, of the test.

Approximately 50% of the questions will relate to a graphic or chart. It is therefore very important to be able to work with material in this format. Once again, do not leave any question blank. Always eliminate answers you know are wrong before you make a guess. If you guess on all fifty questions and get one out of five correct, you will end up with ten correct answers. That is your starting point. When you eliminate answers you know are wrong and pick answers you know are right, your score will go up.

CHAPTER 1 - EARTH SCIENCE

SECTION 1: THE UNIVERSE

The universe is thought to have begun about twenty billion years ago. Because the matter in the universe is expanding (that is, moving outward at fantastic speeds), scientists today favor the **Big Bang** theory to explain how it came into being and evolved into what it is today.

The Big Bang theory states that all the matter was in the same place at one time and a huge explosion took place. The explosion propelled matter outward in all directions. Eventually, the matter developed into millions of billions of galaxies containing millions and billions of stars.

At present, the edge of the universe is thought to be twenty billion light years from earth. (A light year is the **distance** light travels in one year - approximately six trillion miles).

Try a sample question. Treat the material you have just read as a reading passage.

Question 1:

Traveling at the speed of light, how long would it take you to reach the present edge of the universe?

    a. 6 trillion miles
    b. 600 trillion miles
    c. 20 billion miles
    d. 20 billion years
    e. 6 trillion light years

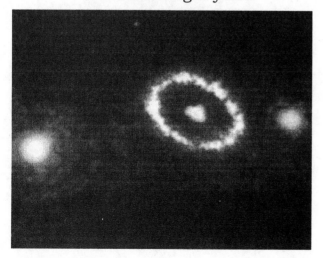

GALAXIES

There are many galaxies in the universe. They vary as to size, shape, and density (amount of matter). Our own galaxy, the **Milky Way**, is between 80,000 to 100,000 light years in diameter and is a two-armed spiral galaxy. It is thickest at its core and thinner in the arms - perhaps having a ten light year thickness.

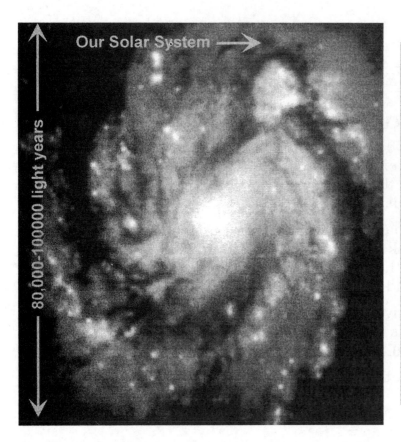

Our Solar System →

80,000-100000 light years

Our solar system is located approximately 2/3 to 3/4 out on one of the arms. It is truly only a "speck of dust" in the total galaxy.

Question 2:

What is the distance across the Milky Way galaxy?

      a. 10 light years
      b. 100,000 light years
      c. 100,000 miles
      d. 6 trillion miles
      e. none of these

THE SOLAR SYSTEM

The sun is our solar system's star. It is an average-sized star approximately five billion years old. It may have come into existence through the collapse of a huge gas and dust cloud called a **nebula**. Compression of the matter, largely hydrogen gas, caused the temperature to rise to a point where fusion could begin. The fusion of hydrogen into helium (and perhaps other types of fusion reactions) creates energy in the form of heat and light. Other forms of electromagnetic energy are created as well. Some of these are radio waves, infra-red rays, U-V, x-rays, gamma rays, and maybe cosmic rays.

Not all of the gas was consumed creating the sun. Some gas formed eddies (whirlpools) in orbit around the sun. These later became the planets, moons (natural satellites), asteroids, and comets.

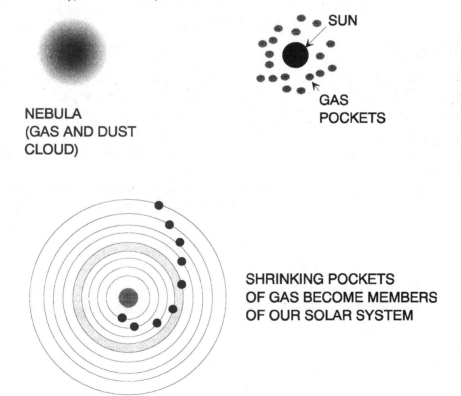

NEBULA
(GAS AND DUST
CLOUD)

SUN

GAS
POCKETS

SHRINKING POCKETS
OF GAS BECOME MEMBERS
OF OUR SOLAR SYSTEM

The solar system is thought to have existed, basically as it is today, about 4.6 billion years ago - that's the age of the earth.

Question 3:

What is not formed from eddies?

        a. planets
        b. moons
        c. comets
        d. asteroids
        e. stars

TERRESTRIAL (SOLID)

INNER PLANETS

THE PLANETS

MERCURY

- The smallest planet.
- Has no atmosphere.
- The surface is marred by meteorite impacts.
- Closest to the sun.
- Has no moons.
- Has a very slow rotation.

VENUS

- Earth's twin (almost the same size).
- Atmosphere is largely CO_2
- Surface is obscured by clouds.
- Rotates east to west, which is the opposite of most planets (retrograde rotation).
- Has a very hot surface temperature (900^0 hot enough to melt lead).
- The third brightest object in the sky after the sun and moon.

Venus's very dense atmosphere is what keeps the planet's surface so warm. The cloud cover warms the surface through the **"greenhouse effect"**.

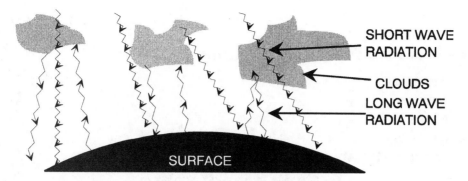

The radiation from the sun has a shorter wavelength, and it is able to pass through the cloud layer. However, after it strikes and warms the surface it is converted to heat. Heat has a longer wavelength and it cannot pass through the clouds as easily as the shorter wavelength radiation. A large portion of the heat is reflected back to the surface by the clouds. The net effect is a warmer surface similar to the way the glass roof warms a greenhouse.

EARTH

- Located just the right distance from the sun for water to exist in all three states of matter (solid, liquid, and gas).
- The only planet with an abundance of water.
- A "live" planet in that it is constantly renewing its crustal material; has weather.
- Has one satellite: the moon. The moon is located 243,000 miles from the earth.
- The moon is the main cause of tides.

MARS

- The "Red Planet"
- Half the size of the Earth.
- Has polar ice caps.
- Has weather (dust storms).
- Shows seasonal changes.
- Has two large moons: Deimos and Phobos.

Question 4:

Which planet is about the same size as the earth?

        a. Mercury
        b. Venus
        c. Mars
        d. Jupiter
        e. None of the above

Question 5:

Which planet is seen first in the evening sky?

        a. Mercury
        b. Venus
        c. Mars
        d. Jupiter
        e. none of the above

THE ASTEROID BELT

The asteroid belt is located between Mars and Jupiter. The asteroids range in size from a grain of sand to 250 miles across (the largest Ceres). There are at least three theories for the formation or the asteroids. The first is that a large planet may have broken apart due to tidal forces. Another theory is that the asteroids are eddies that never combined to form a planet. The last is that they are the remains of two planets that collided and broke apart.

**Gas Giant
Outer Planets**

JUPITER

- The largest planet.
- Has a gigantic red spot in its atmosphere.
- Rotates the fastest of all planets.
- Fast rotation causes the equator to bulge.

SATURN

- Known for its rings
- Second in size.
- Has a large number of satellites.

URANUS AND NEPTUNE

- Very similar in size - third and fourth, respectively.
- Neptune has rings like Saturn.
- Neptune for brief periods of time is the outermost planet.

Question 6:

What is the second largest planet?

        a. Saturn
        b. Uranus
        c. Neptune
        d. Pluto
        e. None of the above

Terrestrial Solid

PLUTO

- The smallest planet.
- Orbits off the plane of orbit of other planets by 17⁰.
- May be a captured planet.
- Usually the farthest planet from the sun.

SECTION 2: THE EARTH

The earth is ninety-three million miles from the sun. After forming, the earth probably underwent changes to produce the earth we know today.

STRUCTURE AND COMPOSITION

Scientists who study earthquakes and meteorites have given us the picture of the interior of the earth. Because the outer core of the earth is liquid, rotational forces produce a magnetic field around the earth. The magnetic poles of the earth wander. At the present time, the magnetic poles are not located where the geographic poles are. They could line up when the magnetic poles wander.

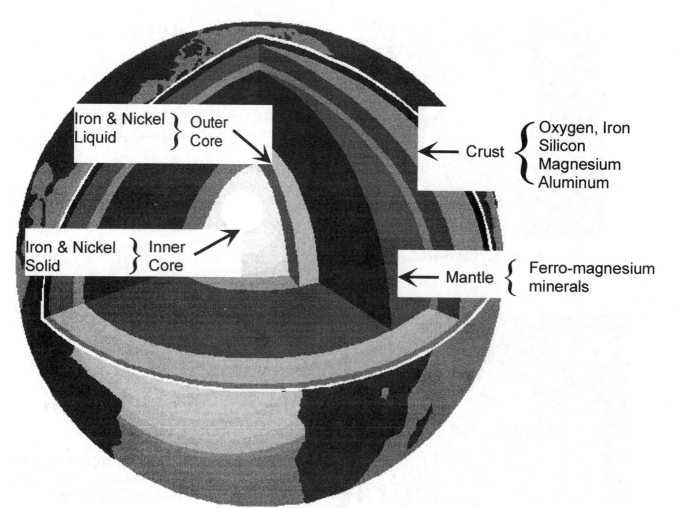

Iron & Nickel Liquid } Outer Core

Iron & Nickel Solid } Inner Core

Crust { Oxygen, Iron Silicon Magnesium Aluminum

Mantle { Ferro-magnesium minerals

Question 7:

What causes the magnetic field of the earth?

        a. magnets in the crust
        b. iron in the crust
        c. the sun
        d. the moon
        e. the rotation of the core of the earth

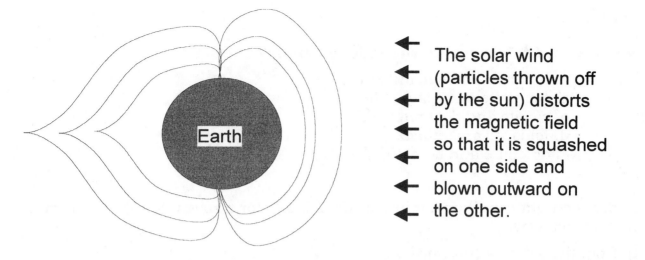

The solar wind (particles thrown off by the sun) distorts the magnetic field so that it is squashed on one side and blown outward on the other.

This magnetic field protects the earth from these solar particles by trapping them in areas around the earth called the **Van Allen Radiation Belts**. These particles are the cause of the northern or southern lights, called the **Aurora Borealis** (N) or the **Aurora Australis** (S).

THE CRUST

The crust is made largely of silicate minerals (over 90%). The composition of the crust may be summarized in the following diagram:

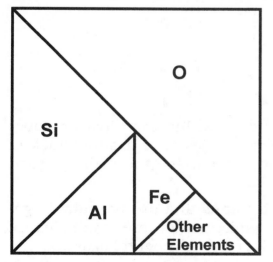

APPROXIMATELY

O = 50%

Si = 25%

Al = 12.5%

Fe = 6.25%

Oxygen and silicon combine to form silica. Under the right circumstances it combines with other metals to form the many silicate minerals that make up the crust. Quartz and feldspar are very common silicate minerals.

Other minerals include:

1) iron ores such as limonite, hematite, and magnetite

2) carbonates such as calcite and dolomite.

15

Question 8:

Which of the following are the most common?

  a. limonite and hematite
  b. hematite and magnetite
  c. calcite and dolomite
  d. feldspar and quartz
  e. dolomite and hematite

Under the right conditions, minerals combine to form rocks. Rocks may form in one of three ways:

1) from melted rock (magma/lava)

2) from cementing together

3) under the influence of heat and pressure.

Question 9:

Which of the following is not a way rocks form?

  a. from melted rock
  b. from crystallization
  c. from cementing
  d. from heat and pressure
  e. all of the above

IGNEOUS ROCK

Igneous rock forms from melted rock material deep within the crust or upper mantle. If the melted rock reaches the surface via a volcano or fissure (crack in crust at surface) it is called **lava**. While below the surface the melt is called **magma** because it is chemically different from lava.

Rocks formed below the surface are called **intrusions** and are typically large-grained because of slow cooling. Granite family rocks are examples of intrusives. Granite type rocks make up the bulk of the continental crust.

Rocks formed on or near the surface are called **extrusions** and are typically very fine-grained or glassy in texture because of rapid cooling. Basaltic family rocks are examples of extrusives. Basalts make up the bulk of the seafloor crust.

Question 10:

Which of the following is true?

    a. extrusives form below the crust
    b. lava and magma are exactly alike
    c. intrusives are small grained
    d. rocks formed by volcanoes are short-grained
    e. granite is a type of extrusive

SEDIMENTARY ROCKS

Sedimentary rocks form from sediments. They are classified as to their composition as well as to how they form.

Clastics - form from particles of other rocks. Typically, a river washes sand, silt, and clay into the ocean where these particles settle to the bottom of the ocean floor. As more particles are deposited on top, water is squeezed out leaving rock "glue" (silica, iron, calcite) holding the particles together and forming a rock. Examples: **conglomerate**, **sandstone**, and **shale**.

Organics - form from the remains of living things. Typically, shells from shellfish deposited in mud (clay) get buried and become fossils. **Chalk**, **fossiliferous limestone**, and **soft coal** (bituminous coal), are examples of organic sedimentary rocks.

Question 11:

Which of the following was never part of a living organism?

    a. shale
    b. coal
    c. diamond
    d. limestone
    e. chalk

Let's continue our review.

Chemicals - form when physical or chemical changes occur in water. There are two ways chemical sedimentary rocks can form:

1) as precipitates - chemical limestone forms this way when temperature changes occur along the ocean landmass interface. CO_2 is given up to the atmosphere as the temperature rises. This allows limestone $(CaCO_3)$[same as calcite] to deposit molecularly on the ocean floor.

2) as evaporates - salt (halite) and gypsum form when bodies of water containing these minerals evaporate, leaving the chemical deposits behind.

Question 12:

How does gypsum form?

    a. by temperature changes
    b. by carbon dioxide being given off
    c. by reaction with the atmosphere
    d. by evaporation
    e. by lava cooling

METAMORPHIC ROCKS

Metamorphic rocks form when rocks (any kind) come in contact with extreme heat (such as liquid rock) or experience tremendous pressures within the crust. Examples of rocks that have changed metamorphically are:

METAMORPHIC

sed. limestone ➜	marble	
sed. shale ➜	slate	
sed. sandstone ➜	quartzite	
ign. granite ➜	gneiss	
meta. slate ➜	phyllite ➜	schist

THE ROCK CYCLE

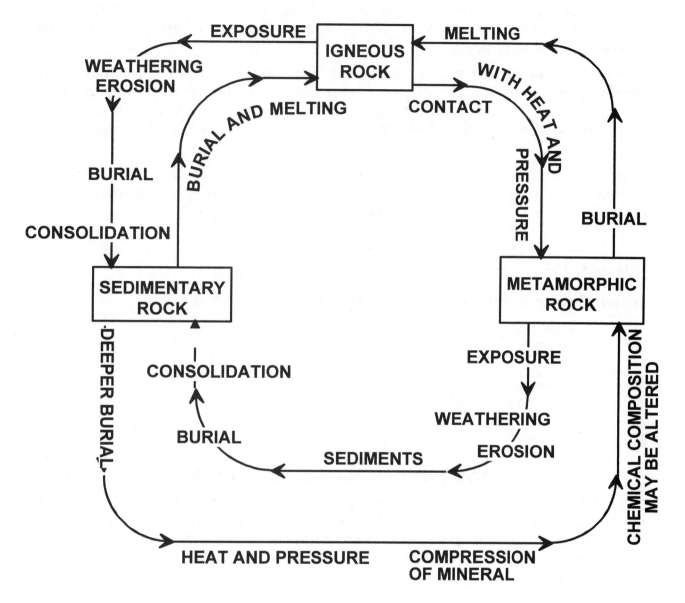

Question 13:

What would have been found in an area before you find marble?

  a. quartzite
  b. limestone
  c. phyllite
  d. slate
  e. gneiss

WEATHERING AND EROSION

The surface of the Earth is constantly changing due to the breakdown and removal of rock material. Weathering, the breakdown of rocks, occurs both physically and chemically.

The work of water in both its liquid and frozen states accounts for most of the physical breakup of rock. Chemically, acids (rain), water, and oxygen do the most damage. Simply, the chemicals combine with rock minerals to corrode and expand the number of cracks or total surface area thereby exposing the rock to more weathering.

Wind, water, and ice are the agents of erosion. Billions of tons of rocks are moved each year by these agents. Glaciers carry all sizes of rocks, while running water carries less, and wind the least.

Question 14:

Which does not have to be removed to prevent erosion?

     a. wind
     b. rain
     c. ice
     d. oxygen
     e. none of the above

SECTION 3: PLATE TECTONICS

After 4.6 billion years of weathering and erosion, the earth should be smooth on the surface. It isn't. Why? Years back, A. Wegener proposed the Continental Drift Theory. He said continents were shaped like jigsaw puzzle pieces and could be fit together. He used South America and Africa as his prime examples.

Today, we are almost sure Wegener's Theory is not true. However, the continents do move - as part of the Earth called the **lithosphere** (a 70-km thick piece of rock - part of which is crust and part upper mantle). The surface is made up of large blocks of rock called **plates**. The blocks are in motion because of the activity of heat and melted rock below the lithosphere in a place we call the **asthenosphere**. Here, partly melted rock moves because of convection, forcing the plates to move away from the mid-ocean ridge area.

Question 15:

What are the large blocks of rock making up the crust float on the mantle commonly called?

     a. plates
     b. continents
     c. blocks
     d. lithosphere
     e. asthenosphere

Mid-Ocean Ridge

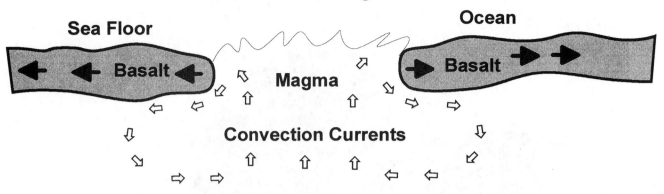

As the plates move <u>away</u> from the mid-ocean ridge, they propel the continents in the direction of movement. We know they are moving because of earthquakes; also, we know there must be some convection because volcanoes are part of the ridge system; in addition sea floor rock is basaltic, formed from lava (melted rock).

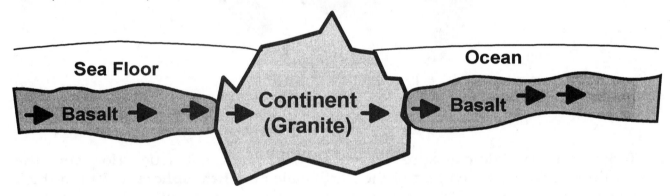

Question 16:

Which of the following are not natural disasters that are caused by moving plates?

        a. volcanoes at the mid ocean ridge
        b. tidal waves
        c. earthquakes
        d. tornadoes
        e. none of the above

Plates may do one of the following three actions:

1) collide

2) separate

3) move beside each other.

Colliding Plates Without Continents

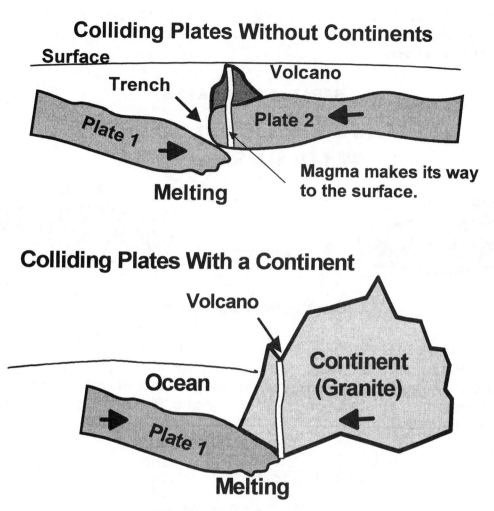

Japan and China are good examples of the above illustration. Japan is volcanic with frequent earthquakes (shallow).

China experiences deep and devastating earthquakes.

Colliding Plates With Two Continents

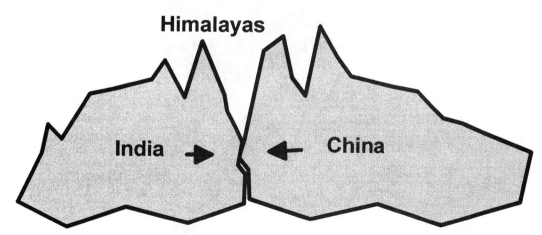

Here, the continents collide, scraping the sea floor up between them and forming very high mountains. Some of the world's large mountain chains are formed in this manner.

SEPARATING PLATES

PLATES MOVING BESIDE EACH OTHER

Mid-Ocean Ridge

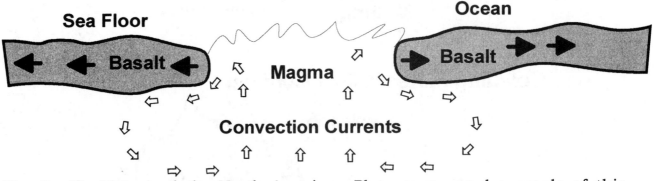

The Pacific Plate and the North American Plate are a good example of this phenomenon. There, most of California (part of North American Plate) and the Southwest part of California (part of the Pacific Plate) slide past each other causing frequent earthquakes.

It is in this way that the earth's surface is constantly renewing itself as crustal rocks are formed at ridges and destroyed at trenches or other subduction (wherever crust moves downward toward the mantle) zones.

Question 17:

What two plates are on either side of the San Andreas Fault?

      a. Mexican and North American
      b. Mexican and Pacific
      c. Pacific and North American
      d. Californian and Pacific
      e. Californian and North American

SECTION 4: THE OCEANS

The oceans cover 71% or three-fourths of the Earth's surface. They average two-and-a-half to three miles in depth and have a salinity (saltiness) of 3.5%. Being liquid, the oceans are always in motion, aiding in the distribution of heat from equatorial regions towards the poles. In fact, the oceans are regarded as the earth's thermostat. Wherever it gets too warm, hurricanes form in tropical waters to provide a quick release of heat energy from that area.

Question 18:

What percentage of the earth's surface is covered by land?

    a. 17
    b. 29
    c. 71
    d. 83
    e. none of the above

SECTION 5: THE ATMOSPHERE

The earth's atmosphere is roughly 600 miles thick. It is composed largely of the free gases, oxygen 21% and nitrogen 78%. Most of these gases are concentrated into the bottom most layer of the atmosphere called the **troposphere**.

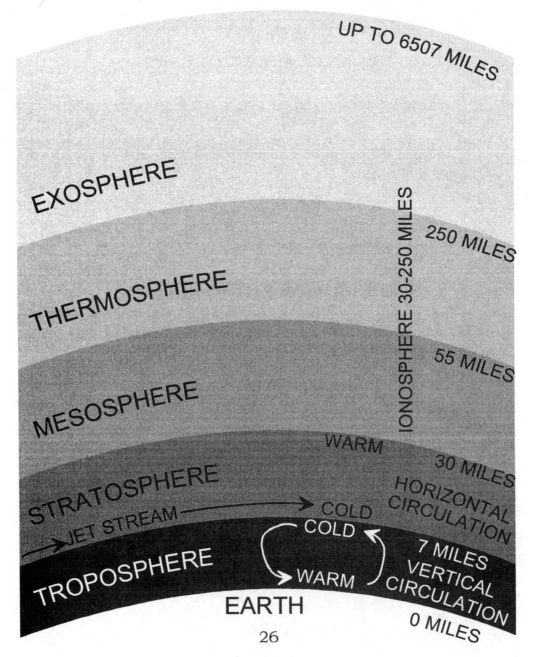

Weather occurs in the troposphere. This layer is warmed as the surface of the earth, rises, cools, and sinks again to replace air that is rising. In other words, convection is operating in the lower atmosphere.

As air rises, it may take water molecules off of the surface. This is called **evaporation**. The water molecules and air move upward and cool. Cooler air cannot hold as much moisture as warmer air. As the gaseous water cools it **condenses** to form clouds. If enough water condenses, the clouds may give up the water again as **precipitation**.

Evaporation, condensation, and precipitation form a cycle known as the **water cycle** or **hydrologic cycle**. This cycle is responsible for the transfer of a large amounts of energy from the surface the atmosphere. Water needs energy just to change from a liquid to a gas - with <u>NO CHANGE IN TEMPERATURE</u>. When the water evaporates, it carries this extra energy up into the atmosphere. When the water condenses, it gives up the energy to the atmosphere. This energy may cause storms to occur.

Question 19:

Which of the following, if stopped, would not break the hydrologic cycle?

     a. rain
     b. evaporation
     c. ice
     d. snow
     e. condensation

Weather is formed both locally and on a broad scale. Locally, thunderstorms are a good example - even monsoon-type weather, although wider in scale, is more dependent upon local geography than, say, polar-front weather systems. Polar fronts are formed when cold, dry air meets warm, moist tropical air. The result is usually precipitation (rain, snow). Temperature, humidity, pressure, and the location of the jet stream may determine the position of polar fronts.

In addition to weather, our atmosphere provides us with a protective layer of poisonous ozone that absorbs ultra-violet light. Ultra-violet light is responsible for sunburn. The ionosphere also reflects certain wavelengths of radio waves allowing for long distance radio communication.

Question 20:

What does ozone protect us from?

     a. poisons
     b. sunburn
     c. radio waves
     d. ionosphere
     e. poisons and sunburn

This completes our review of Earth Science. Before you take the Chapter Test, check your answers for the questions that were found throughout the chapter.

ANSWERS - CHAPTER QUESTIONS EARTH SCIENCE

1. d. The last paragraph states that the edge of the universe is thought to be 20 billion light years from the earth. Therefore if you traveled at the speed of light it would take 20 billion years.

2. b. Be careful when answering this question; the paragraphs state that the galaxy is 100,000 light years across. Make sure you don't pick 100,000 miles.

3. e. The second paragraph states that not all of the gas ended up in the sun. Eddies were formed and when they cooled they formed planets, moons, comets, and asteroids. Stars take the largest percentage of the gas cloud and are therefore not formed from eddies.

4. b. For the answer to this question you must read the paragraphs about the planets. The paragraph about Venus says that it is Earth's <u>twin</u> in size.

5. b. Again you would have to read the paragraphs about the planets. In the paragraph about Venus it says that it is the third brightest object in the sky, behind the sun and the moon. From this information you must make the connection that it would be the first to appear in the evening sky after the moon.

6. a. Reading over the descriptions of the planets, we see that in the paragraph about Saturn it says that it is second in size.

7. e. The paragraph states the rotation of the liquid outer core of the earth causes a magnetic field.

8. d. This question requires you to think a little. Looking at the diagram you can see the silicates are the most common. You must then read to find out which minerals are silicates. This then gives you the most common minerals.

9. b. Crystallization is the one choice that is not listed in the paragraph as a way for rocks to form. Crystallization is how minerals form and then multiple minerals will join together and form a rock.

10. d. The only correct answer is rocks formed by volcanoes are short grained. The second paragraph states that granite is a type of intrusive, and that intrusives are large grained. The first paragraph states that lava and magma are chemically different. The third paragraph states that extrusives form at or near the surface. Rocks formed by volcanoes would form at or near the surface and would consequently be short grained.

11. a. Looking at the paragraph on clastics, we see that shale is the only choice listed. Coal, limestone, and chalk are all listed as organics. Diamond is not listed in either paragraph. Diamond is in fact formed from coal under the influence of heat and pressure. To answer the question you really just had to find a clastic, and since it was never alive it is the correct answer.

12. d. In the paragraph on evaporation, gypsum and salt are the two things listed as examples. Therefore gypsum forms by evaporation.

13. b. Looking at the table of metamorphic rocks, you can see that sedimentary limestone is across from marble. Under the effects of heat and pressure, limestone will turn into marble. Therefore limestone must be found in an area before it can be turned into marble.

14. e. All of the choices: wind, rain, ice, and oxygen, are listed in the paragraphs as being causes of erosion. Wind, rain, and ice are listed as physical erosive agents. Oxygen is listed in the paragraph about chemical agents.

15. a. The last paragraph states the large blocks of rock floating on the mantle are called plates. Always read carefully, many people will mistakenly choose continents for this question.

16. d. Volcanoes and earthquakes are all listed as proof of the plates' movement. Tornadoes, being above ground, would not be affected by plate movement. Looking closer at answer b, tidal waves occur when an earthquake happens underwater.

17. c. For this question you have to read the paragraphs until you find a reference to the San Andreas Fault. It is found in the last paragraph. It states the San Andreas Fault is caused by the collision of the Pacific and North American plates.

18. b. This is a trick question. The paragraph states that 71% of the Earth's surface is covered by water. If you answer 71% you are wrong because the question asks for how much of the Earth is covered by land. The amount covered by land would be whatever is not covered by water, or 29%.

19. c. Rain and snow are both types of precipitation and would be part of the hydrologic cycle. Evaporation and condensation are both listed as parts of the hydrologic cycle. The only choice left, ice, is not part of the hydrologic cycle. The formation of ice on lakes is separate from the cycle that forms the weather.

20. b. The second paragraph states ozone absorbs sunburn-causing ultraviolet light. It also says ozone is poisonous; it does not say it protects us from poison.

After you have reviewed the areas where you had difficulty, you will be ready to take the Earth Science Chapter Test that begins on the next page.

As stated in the introduction, when you take the GED SCIENCE test you will be given a passage to read or a graph or chart to study. From what is directly stated or you are able to infer from the passage, you will be expected to select the correct answer for each question from the five possible choices. This will give you valuable practice in choosing the correct answer from what you read, which is the skill you need to develop in order to achieve on the GED SCIENCE TEST. Remember, there should only be about eleven questions on the test concerning space and earth science.

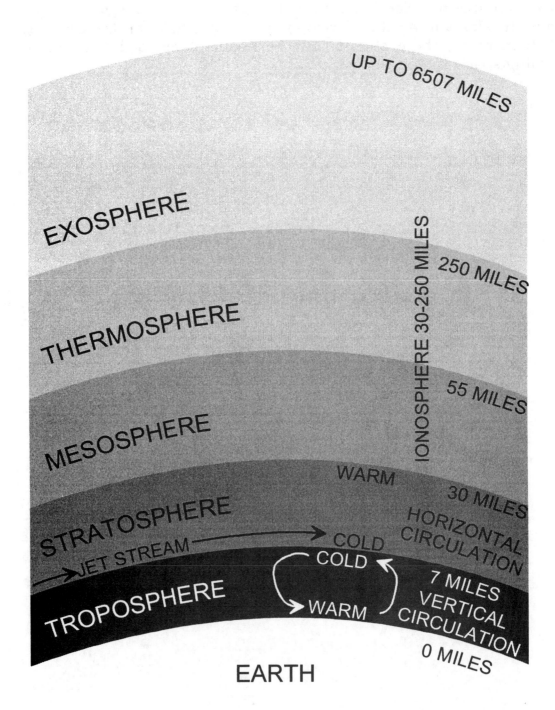

1. **The bottom layer of the atmosphere is called the**
    a. tropospanse
    b. troposphere
    c. ionosphere
    d. exosphere
    e. stratosphere

2. **You call a radio station to complain about the poor reception you have been getting, and they respond that there has been a big disruption in the ionosphere. At what height did this disruption occur?**
    a. 1 mile
    b. between 1 and 7 miles
    c. between 7 and 15 miles
    d. between 7 and 30 miles
    e. over 30 miles

3. **Limestone is formed when millions of sea animals die and their shells get compressed into a rock. You believe that marble then forms when limestone is under high temperature and pressure. Which of the following if true would help back up your assumption?**
    a. volcanoes at the mid-ocean ridge.
    b. marble is often found at the site of ancient seabeds.
    c. when you dig marble at the quarry it is warm.
    d. marble is easy to polish.
    e. marble is found at the site of an ancient glacial deposit.

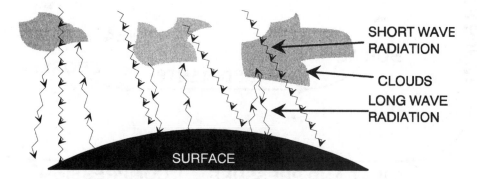

4. **What is a contributing factor to the greenhouse effect?**
    a. the surface of the planet warms up
    b. the changes in solar radiation
    c. short-wave radiation passes through clouds easier than long-wave radiation.
    d. the water cycle
    e. evaporation

5. **Planets are formed when eddies around a newly formed star cool. Which of the following would have formed from the largest eddy?**

    a. Pluto
    b. Jupiter
    c. Mars
    d. Earth
    e. Mercury

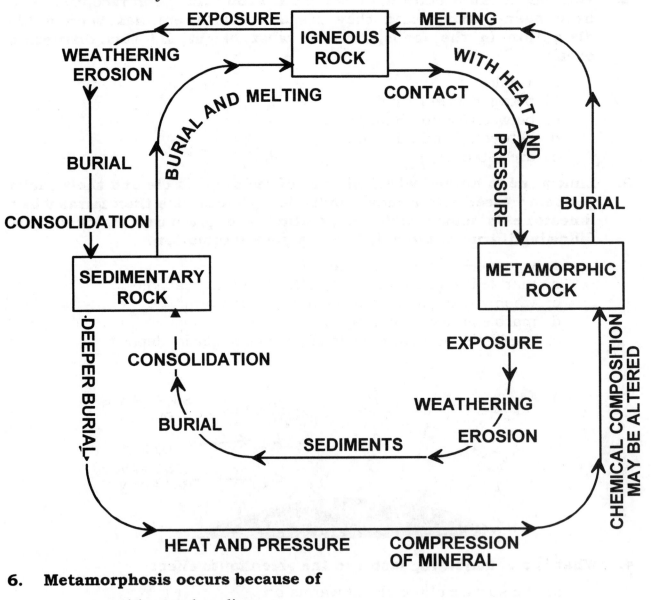

6. **Metamorphosis occurs because of**

    a. melting and cooling
    b. erosion and weathering
    c. erosion and deposition
    d. burial and melting
    e. extreme heat and pressure short of total melting

7. **According to the theory of plate tectonics the surface of the earth is constantly being renewed and reformed. The North American and the European continents are moving farther apart. At this spread a ridge of volcanoes adds extra material. Every action has an opposite reaction. In other places the plates are colliding and one plate is being pushed under the other. In the theory of Plate Tectonics, plates are renewed at**

    a. ridge areas
    b. trenches
    c. volcanoes
    d. fault zones
    e. continents

8. **Large mountain chains form when**

    a. plates slide past each other
    b. plates separate
    c. oceanic plates collide
    d. two continents riding on plates collide
    e. volcanoes form

9. **Of the following, which one is part of the hydrologic cycle?**

    a. maturation
    b. inebriation
    c. condensation
    d. saturation
    e. wind

10. **Halite is found in the remains of old seabeds. Over time the seas dried up and halite was left behind. Which of the following best describe halite.**

    a. metamorphic rock
    b. sedimentary clastic rock
    c. sedimentary chemical evaporite rock
    d. sedimentary organic rock
    e. igneous rock

11. **The moon or sun pulling on the ocean causes tides. Even though the sun is much larger than the moon, because the moon is so much closer, the moon has more of an effect on the tides. What do you think would cause the largest tides?**

    a. The sun and the moon pulling in the same direction
    b. The sun pulling directly opposite the moon
    c. The sun pulling at a 90^0 angle from the moon
    d. The moon pulling on a small body of water.
    e. The sun pulling on a small body of water.

ANSWERS - CHAPTER TEST EARTH SCIENCE

1) b Looking at the diagram, you need to find the lowest level. This level is the troposphere and it goes from zero to seven miles in height. This is the level where our weather occurs.

2) e The ionosphere goes from thirty to 250 miles in height.

3) b If marble is found in areas where ancient seas were found, this would back up your assumption. Volcanoes would have nothing to do with your assumption. Whether marble is warm when you dig it would not prove or disprove your theory because the heat and pressure transformation take place over time. The fact marble polishes easily has nothing to do with its formation; it is simply one of the physical properties of marble.

4) c Choice a is the net result of the greenhouse effect. The best choice is c because it gives the exact reason why the radiation comes in through the clouds, but has a harder time getting back out.

5) b To answer this question you will need a small amount of previous knowledge, that Jupiter is the largest planet. The largest planet would have the largest eddy.

6) e Metamorphosis occurs when rock under high heat and temperature changes to a different form. However, the temperature must remain lower than the melting point, or the rock will melt instead of going through metamorphosis. Erosion does not change the type of rock; it simply breaks it into smaller pieces.

7) a The plates are renewed at the ridge areas. Ridges are a long series of volcanoes that spew out new material. The answer is not simply volcanoes because the question asks for a location (ridge areas) not how (volcanoes).

8) d Let's look at each answer individually. Plates sliding past each other do not make huge mountains because most of the force is spent overcoming friction. When plates separate, there is, again, no force to create a mountain. When oceanic plate collide, one can slide under the other forming a trench or they can crumple up for islands. However, they started so low below the surface the do not appear as a large mountain chain. Two continents colliding is the best answer. Here the force has nowhere to go but into forming mountains. The plates crumple up as they crash into each other. Not all mountain ranges are volcanic in nature.

9) c The hydrologic cycle starts with water evaporation from the surface. The water then condenses in clouds and finally it precipitates back to the Earth.

10) c The key here is to notice the words "dried up." This lets you know that evaporation is occurring. There is no mention whether halite is able to undergo metamorphosis or that it is made from living material.

11) a The best choice is clearly when the sun and moon are pulling in the same direction. Here their forces will be added together. When you take a look at tides, the moon (and the sun) not only pull the ocean close to them up away from the earth, but they also pull the earth away from the ocean on the other side. This causes the tide to rise on the side of the earth opposite the moon (or sun). When the sun and moon pull in opposite directions there will also be large tides but some of their forces will cancel each other out. When the sun and moon pull at 90^0 the tides are smaller because the sun and moon are canceling each other out. When the sun and the moon pull on small bodies of water they do not have as large of an effect because there is no more water to flow into the area to raise the water level.

SECTION 1: THE CELL

In order for us to understand how living things function, we must first understand the basic unit of life. Every living thing we see, both plants and animals, is made up of the same basic building block, the cell. There are differences between cells, but it is surprising how similar most cells are to each other.

Let's take a look at what a typical animal cell looks like so you can get an idea of what some of the basic parts do, and what they look like.

ANIMAL CELL

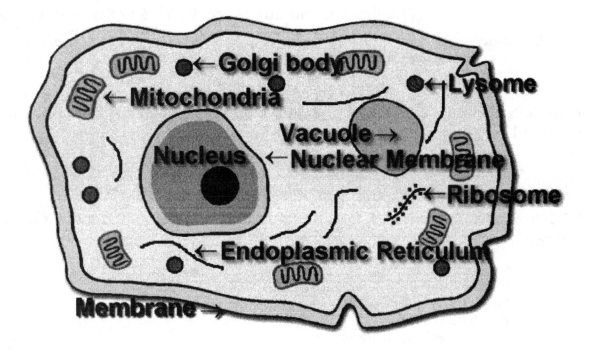

1. **Centrioles -**   Aid in spindle formation during cell division. It makes sure that the chromosomes (the genetic material) go to the right place.

2. **Endoplasmic Reticulum** - A structure used to separate areas of the cell. It looks much like the cell membrane.

3. **Cell Membrane** - Holds the cell together. It also has the job of letting things in and out of the cell. It is very thin, and goes all around the outside of the cell.

4. **Nuclear Membrane** - Separates nucleus from rest of cell body.

5. **Chromatin** - the "genetic" material of the cell. It is responsible for determining what the cell does.

6. **Golgi Body** - stores proteins.

7. **Ribosome** - site where protein synthesis occurs.

8. **Nucleolus** - supplies M-RNA chemicals for protein synthesis.

9. **Mitochondria** - site where energy for cell functions is released.

10. **Cytoplasm** - a gel-like substance containing the following: water, minerals, vitamins, proteins (enzymes), carbohydrates, fats (lipids), and nucleic acids. This is the fluid in a cell.

11. **Vacuole** - a space within the cell that stores food, water, or waste material.

12. **Lysome** - aids in the breakdown of proteins.

Now, let's try some questions. The answers are found in the material you have just read.

Question 1:

Which of the following if missing would cause the cell to die for lack of energy?

      a. the vacuole
      b. the ribosome
      c. the cell membrane
      d. the nucleus
      e. the mitochondria

Question 2:

The main function of the cell membrane is

      a. protein synthesis.
      b. regulation of things entering and leaving the cell.
      c. energy production.
      d. cell reproduction.
      e. storage.

SOME THINGS TO REMEMBER:

1.   Chromatin forms chromosomes during reproduction.

2.   The cell membrane (nuclear membrane) allows substances to pass through - water, gases, food, waste, etc.

3.   The nucleus controls metabolism, growth, and reproduction.

The following illustration points out the major differences between the plant and the animal cell. Most of the parts remain the same, but plants have several extra parts because they do different jobs than animal cells.

PLANT CELL

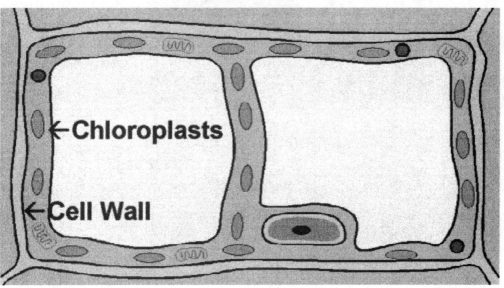

Plant cells usually do not have the mobility of animal cells. Thus, they have a need for protection from their surrounding environment. This protection comes in the form of a non-living cell wall made of cellulose. This is a hard material. A common form we know is wood.

The **chloroplasts** in plant cells manufacture their own food. They are green in color and contain chlorophyll, which is used to collect energy from the sun. This energy is used to help make food for the plant cell.

Question 3:

The cell wall in a plant is

  a.   living and made of cellulose.
  b.   living and made of chlorophyll.
  c.   living and made of wood.
  d.   dead and made of cellulose.
  e.   dead and made of chlorophyll.

Question 4:

The main function of the cell wall is

      a. protection.
      b. movement.
      c. energy production.
      d. photosynthesis.
      e. cell reproduction.

SECTION 2: HOW PLANTS GET THEIR FOOD

The food making process in green plant cells is called **photosynthesis.**

The following equation explains what occurs during the photosynthesis reaction.

$$\text{light energy} + 6CO_2 + 6H_2O \xrightarrow[\text{+ enzymes}]{\text{Chlorophyll}} C_6H_{12}O_6 + 6O_2$$

The light energy changes to chemical energy (chloroplasts). The chemical energy is used to unite carbon dioxide (CO_2) with water (H_2O) to produce sugar ($C_6H_{12}O_6$) and oxygen (O_2), which is released as a gas.

From this process of photosynthesis, all the food for all the plants and animals is produced. Plants make food and grow; animals then eat these plants, which gives them food.

Question 5:

During photosynthesis, what two molecules are combined?

  a. water and sugar
  b. sugar and oxygen
  c. water and carbon dioxide
  d. oxygen and carbon dioxide
  e. water and oxygen

Question 6:

If it were not for photosynthesis,

  a. all the plants would die.
  b. all the animals would die.
  c. animals would take over the earth.
  d. both plants and animals would die.
  e. plants would take over the earth.

A plant cell that does not make its own food may be **parasitic** (living off other plants or animals) or **saprophytic** (Living off dead plants or animals).

OTHER POINTS TO REMEMBER:

- Animal cells capture their food. They do not make it. They must eat and then break down their food to get energy.

- Any form of energy may be changed to any other form. In fact, this must regularly be done if the organism is going to survive.

Question 7:

Which of the following is true?

  a. Animals can make their own food.
  b. Fish can make their own food.
  c. All plants make their own food.
  d. Organisms cannot change molecules.
  e. Some plants can not make their own food.

SECTION 3: DNA, RNA AND PROTEIN SYNTHESIS

DNA (deoxyribonucleic acid) makes up the genes that form the chromosomes. This is very important because the genes carry all the information to build and run the organism.

A set of chromosomes contains all the information needed for a cell to survive and reproduce. If a nucleus, which contains the information, is removed from the cell, the cell dies.

During reproduction, the chromosomes are duplicated and passed along to the next generation. This means that when a cell divides, both resulting cells have exactly the same genetic material.

This is why a lizard always looks like a lizard and a cat looks like a cat. One of the most important things for living organisms to do is to pass along their information to their offspring.

Question 8:

DNA stands for

> a. deoxyribonucleic acid.
> b. dextrose near arrangement.
> c. doesn't need alcohol.
> d. denitro acid.
> e. denatured alcohol.

Question 9:

Which of the following is not true?

> a. Chromosomes are duplicated during cell reproduction.
> b. Animal cells have chromosomes.
> c. Plant cells have chromosomes.
> d. Cells need chromosomes to live.
> e. Chromosomes grow continuously.

Chromosomes are genes that are joined together. Genes are made up of set combinations of DNA molecules. These set combinations are the cell's genetic information.

44

A DNA molecule's shape looks like a twisted ladder. The following diagram should give you a better idea of what it looks like.

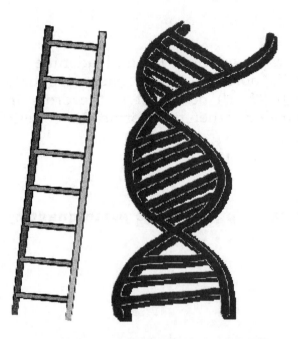

This formation in a molecule is called a **double helix**. The outside supports are made of sugars and phosphates, while the rungs (stem) are made of nucleic bases.

LADDER   TWISTED LADDER

Question 10:

The rungs of the ladder in a DNA molecule are

    a. made of sugars.
    b. made of nucleic bases.
    c. broken.
    d. shifting constantly.
    e. made of phosphates.

QUESTION 11:

Genes

    a. make up molecules.
    b. float freely in the cytoplasm.
    c. are linked on chromosomes.
    d. make proteins.
    e. are found outside the cell.

The four types of bases are:

1. cytosine
2. guanine
3. adenine
4. thymine

Cytosine (1) always pairs with guanine (2) and adenine (3) with thymine (4) in a DNA molecule. These bases are held together by weak hydrogen (H—H) bonds. It is the order in which these bases are joined together that forms the genetic code.

Question 12:

In a DNA molecule, which of the following is found in the same quantity as adenine?

a. cytosine
b. guanine
c. uracil
d. thymine
e. none of these

G
C
U
A
A
G
U
C
C
A
U

RNA (RNA - ribonucleic acid) is patterned after DNA. Actually, it is copied from the DNA. Just as many people backup important computer files to protect them, the cell copies the DNA molecule so there is less chance of it getting damaged. The chemical difference between RNA and DNA is **uracil**, which replaces the thymine.

RNA can pass through the nuclear membrane and enter the main part of the cell, DNA canno...

Question 13:

What can RNA do that DNA can't?

a. duplicate itself
b. leave the nucleus
c. divide the cell
d. form a double helix
e. store genes

In the following example we will follow the transfer of information from the nucleus (chromatin/gene matter) to somewhere inside the cell body.

HOW THE NUCLEUS TALKS TO THE CELL

MESSAGE: forms a special enzyme (protein) for energy production

STEP I: DNA in the nucleus contains the original genetic message.

Straightened DNA Chain

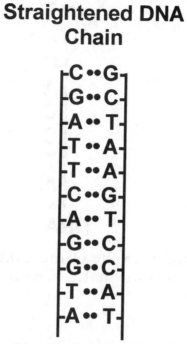

•• **Hydrogen Bonds**

STEP 2: The molecule begins to separate at the Hydrogen bonds.

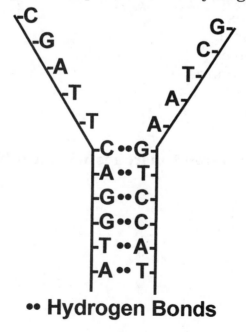

•• **Hydrogen Bonds**

STEP 3:   As the separation occurs, RNA bases begin to link with the "message" base combination.

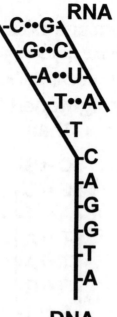

Notice cytosine, guanine, and thymine of the DNA strand all match with their complementary base, while adenine matches with a new base-uracil.

Question 14:

Where is the RNA message formed?

        a. In the cytoplasm.
        b. In the nucleus.
        c. In the ribosome.
        d. In the centriole.
        e. In the mitochondria.

Question 15:

Which base in the DNA strand matches with a different base in the RNA strand?

        a. adenine
        b. thymine
        c. guanine
        d. cytosine
        e. uracil

STEP 4: The RNA strand now has the "message" and leaves the nucleus for the cell body. The DNA continues to make these RNAs until stopped by some chemical stimulus. These RNAs are called **Messenger RNA** or **M-RNA**.

STEP 5: M-RNAs eventually end in the area of the ribosomes. Ribosomes contain ribosomal RNA and a sort of mutual attraction between it and the M-RNA occurs.

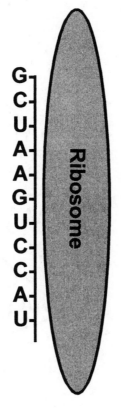

Question 16:

When the M-RNA leaves the nucleus, where does it end up?

        a. It continues out of the cell.
        b. It goes to the mitochondria.
        c. It goes to the ribosome.
        d. It goes to vacuole.
        e. It goes to the Golgi body.

Question 17:

What stops the DNA from making more M-RNA?

        a. the nucleus
        b. the ribosome
        c. the cell body
        d. the ribosomal RNA
        e. a chemical stimulus

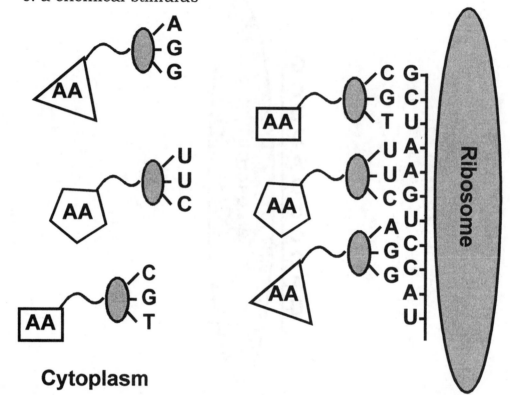

Cytoplasm

Floating out in the cytoplasm are nucleic acids, some of these nucleic acids are transfer RNAs with amino acids (amino acids make up proteins, or enzymes) attached to them. There are twenty different types of amino acids.

STEP 6:   The **Transfer RNAs** (T-RNA) come in contact with the M-RNAs attached to the ribosome. When the right combination matches the M-RNA, they "stick" together for a while until the process completes itself.

Question 18:

How many different amino acids are there?

        a. 5
        b. 7
        c. 13
        d. 20
        e. 27

Question 19:

What type of RNA has amino acids attached to it?

     a. messenger
     b. amino acid
     c. transfer
     d. reflux
     e. chromosomal

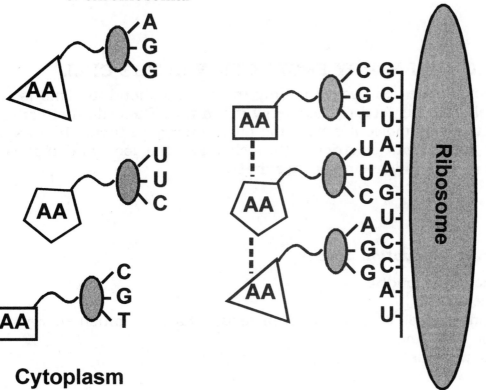

Cytoplasm

As you can see above, there are many combinations of T-RNAs - all carrying a specific amino acid.

The T-RNAs attach to the M-RNA. The amino acids combine together to form an enzyme (protein).

The enzyme will eventually encounter an energy-making situation and do its job. The T-RNA, now minus their specific amino acids, are released back into general circulation in the cytoplasm where they will eventually find their complementary amino acids. The M-RNA continues to attract T-RNAs until it is chemically shut down.

Question 20:

The T-RNA can

       a. attach to any amino acid.
       b. attach to any point on the M-RNA.
       c. attach to a specific amino acid.
       d. leave the cell.
       e. form a protein.

SECTION 4: ENERGY PRODUCTION IN THE CELL

Energy production in the cell occurs mainly in the mitochondria. This is the site where the cell converts food to the energy it can use. The cell must convert the energy, because food molecules have too much energy in them. Just as we convert nuclear energy into electricity, the cell takes the food and makes a chemical known as **ATP** (adenosine triphosphate).

Question 21:

Where does energy production occur?

       a. In the mitochondria.
       b. At the ATP.
       c. Outside the cell because sugar molecules have too much energy.
       d. At the ribosome.
       e. In the nucleus.

Below is a chemical reaction. It shows how a sugar molecule is broken down. Notice that it is exactly the opposite reaction we saw in photosynthesis. Here energy is being released instead of being stored in the sugar molecule.

OXIDATION

$$C_6H_{12}O_6 + 6O_2 \quad \xrightarrow{\text{ENZYMES}} \quad 6CO_2 + 6H_2O + energy$$

Question 22:

Oxidation is the opposite of what other reaction?

    a. enzymea
    b. photosynthesis
    c. electrolysis
    d. hydrolysis
    e. genesis

In hydrolysis, water combines with a fat to break the fat down into fatty acid and glycerol. The reaction releases a small amount of energy. However, the glycerol can then enter the oxidation pathway that the sugar molecule followed. This allows the cell to get more energy from the fat.

HYDROLYSIS

H_2O + fat → fatty acid + glycerol + small amount of energy

Question 23:

What two chemicals react in hydrolysis?

    a. fat and fatty acid
    b. fat and water
    c. water and fatty acid
    d. water and glycerol
    e. fatty acid and glycerol

Now, let's take a look at how the cell uses the energy given off in these reactions. Remember the cell must be able to convert this energy to small units that it is able to use. There are two main chemicals involved.

ATP - adenosine triphosphate (carries three phosphate molecules)

ADP - adenosine diphosphate (carries two phosphate molecules)

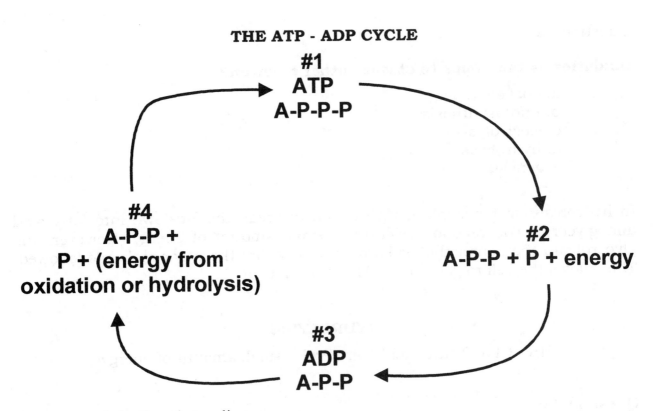

Let me explain the above diagram.

STEP 1: This is an ATP molecule. It has energy stored in it for the cell. When the cell does not need energy it just floats around.

STEP 2: The cell now needs energy. It takes the ATP molecule and breaks off one of the phosphates. This releases energy for the cell to use.

STEP 3: The cell has used the ATP molecule for energy, but has enough other ATP molecules around so it does not have to make more by breaking down food molecules.

STEP 4: The cell supply of ATP is getting low. It is time for it to make more. It breaks down food to release energy. It then takes that energy to attach a phosphate to an ADP molecule. We are now back at the ATP molecule in Step 1.

Question 24:

When the cell's supply of ATP is high, which of the following is true?

       a. The cell converts ATP to ADP.
       b. The cell converts ADP to ATP.
       c. The cell stops converting ADP to ATP.
       d. The cell breaks down food.
       e. The cell stops making ADP.

Question 25:

When the cell needs small amounts of energy, which of the following is true?

    a. The cell converts ATP to ADP.
    b. The cell converts ADP to ATP.
    c. The cell stops converting ADP to ATP.
    d. The cell breaks down food.
    e. The cell stops making ADP.

SECTION 5: CELL DIVISION - MITOSIS

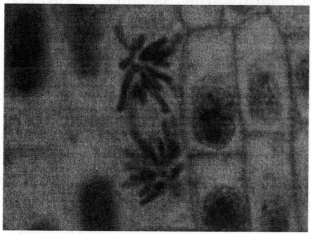

The process of a cell making an <u>exact copy</u> of itself is called **mitosis**. This is a very important process, because both cells (the original and the copy) must end up having the <u>same</u> information. If the cells do not get the same information, one of the cells will probably die because it does not have everything it needs.

Plant cell undergoing mitosis

Question 26:

Why must the cells end up with the same information?

    a. because they are supposed to
    b. because they must reproduce
    c. because they do not need to get the same material
    d. because one cell would likely die if it did not get the right material
    e. because every cell has different genetic material

The following are the basic steps in mitosis, identical cell reproduction:

ANIMAL CELL

STEP 1: **Interphase**

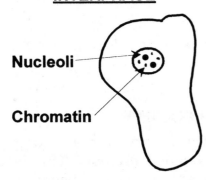

INTERPHASE

Nucleoli

Chromatin

This is the time between the last cell division and immediately before mitosis is ready to begin again. The cell must grow. After the last division, the cell became half its original size. Now it is growing. The **chromosomes**, with all the cell's genetic material in the form of DNA, are unwound and are spread throughout the nucleus. In this form they are simply called **chromatin** because individual chromosomes are not visible. It is necessary for the chromosomes to unwind for them to be able to make RNA. The final step in this phase is when the DNA makes a copy of itself. The cell now has two copies of the DNA so it can divide and give a set to each cell.

Question 27:

Which of the following does not occur during interphase?

    a. the cell divides
    b. the cell duplicates its DNA
    c. the cell grows
    d. the cell chromosomes unravel
    e. the cell makes RNA

STEP 2: Mitosis Begins Prophase (Animal Cell)

PROPHASE

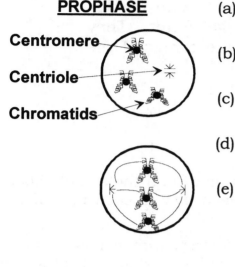

Centromere

Centriole

Chromatids

(a) The centriole divides. Each half then migrates to a different half of the cell.

(b) Fibers start to radiate from the centriole halves. They are called **aster** fibers.

(c) The chromatin in the nucleus starts to coil up into its chromosome phase.

(d) The nucleoli, one of the structures in the nucleus, disappear.

(e) The chromosomes continue to coil. They are now called **chromatids**. The identical chromosomes formed during interphase join so you have pairs of chromosomes. The pairs are joined together at their center by a **centromere**.

(f) The nuclear membrane disappears.

(g) The chromatid pairs migrate toward the equator and line up down the middle of the cell. The **aster** fibers have gotten longer and they are now called the **spindle**. One spindle fiber from each half of the cell attaches to one of the chromatids in each pairing of chromosomes.

Question 28:

Which of the following does <u>not</u> occur during prophase?

    a. The chromosomes coil.
    b. The nuclear membrane disappears.
    c. The centriole splits.
    d. The aster fibers form.
    e. The cell membrane disappears.

Question 29:

What do the aster fibers attach to?

    a. the nuclear membrane
    b. the cell membrane
    c. the chromosomes
    d. the nucleoli
    e. the ribosomes

STEP 3: Metaphase

METAPHASE

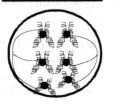

During metaphase the **chromatids** continue to shorten and get fatter. They are now in a line down the center of the equator.

STEP 4: Anaphase

ANAPHASE

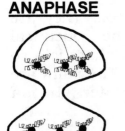

(a) The chromosome pairs break apart and the spindle fibers pull one chromosome from each pair to each pole of the cell.

ANIMAL 1. ANIMAL CELL - A cleavage furrow pushes in at the middle of the cell and slowly divides the daughter cells.

PLANT 2. PLANT CELL - A cell plate forms down the middle of the cell and it divides the daughter cells.

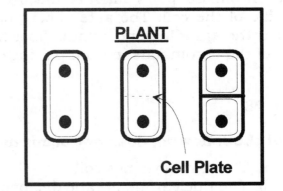

ANIMAL

Furrow

PLANT

Cell Plate

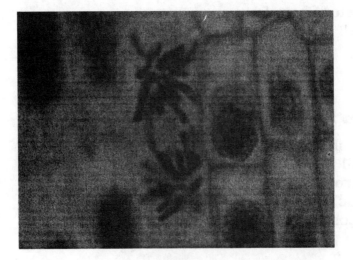

Question 30:

What is the difference between plant cell division and animal cell division?

        a. A plant cell develops a cleavage furrow.
        b. There are no differences.
        c. A plant cell forms a cell plate.
        d. Plant cells end up with eight cells.
        e. Animal cells end up with four cells.

STEP 5: Telophase

Telophase is the opposite of prophase in that the fibers disappear and the nucleus reforms. It contains both nucleoli and chromatin.

Cell division is the same for plant cells, except plants do not have centromeres. The spindle fibers form without them.

Question 31:

Telophase is the opposite of what other stage of mitosis?

        a. interphase
        b. prophase
        c. metaphase
        d. anaphase
        e. none of these

SECTION 6: CELL DIVISION - MEIOSIS

Meiosis is sex cell formation. In meiosis, cell division occurs in two parts. The first part is called the first meiotic division. The cell divides and splits all the genetic information in the same fashion as mitosis. This leaves the cell with two copies of each chromosome.

The second part involves the reduction of the chromosome pairs into single chromosomes. This leaves each cell with only one copy of each chromosome. This is very important because if the organism could not reduce the number of chromosomes in its sex cells, the number of chromosomes would always increase when it mated. To put this another way, if an organism got two of each chromosome from each parent it would end up with four of each chromosome. A cell with only one set of chromosomes is called **haploid**. A cell with two sets of chromosomes is **diploid**.

The resulting cells in a male organism are sperms. Four sperms are produced for each original cell. They all are able to fertilize an egg.

In a female, the resulting cells are the öotid (egg) and polar bodies. When the cells in a female divide, all of the cytoplasm goes with one cell. This becomes the egg. The other cells are called polar bodies. They have very little cytoplasm and quickly die.

If fertilization occurs, a sperm containing a haploid number of chromosomes unites with an ovum containing a haploid number of chromosomes. The fertilized cell, a **zygote**, contains a diploid number of chromosomes. In the diploid state, regular mitotic division can occur and the zygote becomes the kind of living being dictated by the genes.

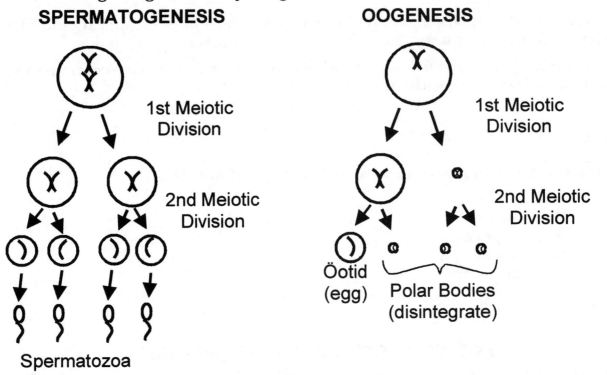

SPERMATOGENESIS

OOGENESIS

1st Meiotic Division

2nd Meiotic Division

Spermatozoa

1st Meiotic Division

2nd Meiotic Division

Öotid (egg)   Polar Bodies (disintegrate)

Question 32:

What is a haploid cell?

>   a. a cell that can only move sideways
>   b. a cell that is ready to divide
>   c. a cell that is in the process of division
>   d. a cell that has only one set of chromosomes
>   e. a cell that has just finished dividing

Question 33:

Which of the following is true?

    a. four sperm cells are produced in meiosis
    b. four egg cells are produced in meiosis
    c. three of the four sperm cells die quickly
    d. only one sperm cell is viable
    e. egg cells are diploid

SECTION 7: GENETICS

Gregor Mendel is called the "father of genetics". In the early 1900's discoveries and conclusions gathered from his work with garden peas were published.

Question 34:

Another name for Gregor Mendel is?

    a. the father of genetics
    b. Father Time
    c. the father of our country
    d. Father John
    e. Father Garden Pea

The following is a summary of the basic laws he discovered.

THE MENDELIAN LAWS OF INHERITANCE

1. The Law of Dominance   Each gene has two copies, one from the mother and one from the father. One of the copies may be **dominant** and show its characteristic over the other copy. If so, the other copy is considered to be **recessive**.

EXAMPLE: A pure tall pea plant crossed with a pure short pea plant produced seeds that produced only tall pea plants.

CONCLUSION: The "tall" factor (gene) was dominant over the "short" factor (gene).

TALL/DOMINANT

SHORT/RECESSIVE

Question 35:

Which of the following is not true?

    a. One gene is always dominant.
    b. A dominant gene masks a recessive gene.
    c. Most organisms have two copies of most genes.
    d. An organism has a set number of genes.
    e. Genes are found on the chromosomes.

Traits are represented by letters to quickly identify different characteristics. Capitals are employed to note dominance and lower case to show recessiveness.

EXAMPLE: (a) TT Gene pair indicating "tall" dominance

          (b) tt Gene pair indicating "short" recessiveness

          (c) Tt Gene pair indicating that the "tall" gene will dominate the recessive "short" gene

Gene pairs are often referred to as being either **homozygous** (genes with similar traits like (a) and (b) above) or **heterozygous** (genes with different traits, such as (c) above). The letters (TT, tt, or Tt) are referred to as the **genotype**, the genetic composition, of the organism.

Question 36:

What is a heterozygous animal?

       a. an animal with two alike genes
       b. an animal with only one set of chromosomes
       c. an animal with four genes
       d. an animal with unlike genes
       e. an animal with two recessive genes

The trait, which determines the physical appearance (tall, short), is called the **phenotype**.

EXAMPLE: **A CROSS BETWEEN A PURE TAN MOUSE AND A PURE WHITE MOUSE**

Phenotype	Homozygous Pure Tan	Homozygous Pure White
Genotype	TT	tt
Genotype	Tt Tt Tt Tt	
Phenotype	All hybrid or heterozygous tan	

CONTINUING, A CROSS BETWEEN TWO HYBRID TAN OFFSPRING

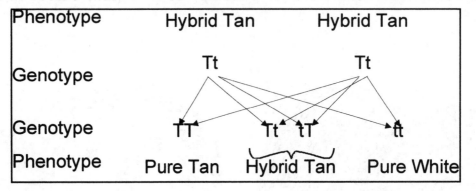

Phenotype	Hybrid Tan	Hybrid Tan
Genotype	Tt	Tt
Genotype	TT Tt tT tt	
Phenotype	Pure Tan Hybrid Tan Pure White	

Question 37:

Which of the following is true?

     a. Hybrid tan mice (Tt) will only have white children.
     b. Hybrid tan mice (Tt) will only have tan children.
     c. Hybrid tan mice (Tt) will have white and tan children.
     d. Hybrid tan mice (Tt) will only have children of one phenotype.
     e. Hybrid tan mice (Tt) will only have children of one genotype.

2. The Law of Segregation   Mendel concluded the genes were separated during the formation of the sex cells (meiosis) and they (the genes) recombine during fertilization.

3. The Law of Independent Assortment   Mendel concluded one trait might be inherited independently from another trait.

EXAMPLE: Tall pea plants may produce either wrinkled or smooth seeds (peas). Short plants showed the same diversity.

Question 38:

Mendel concluded that

> a. pea plants are very unique.
> b. all genes are linked.
> c. genes are mixed during mitosis.
> d. genes are separated during mitosis.
> e. genes can be inherited independently.

SECTION 8: LIVING THINGS

The classification of living things has long been the subject of debate. Previously, living things were classified as being plant or animal. Improvement in microscopes, better chemical analysis of organisms, and the wide diversity of living things, has led most scientists to agree that there are no less than three major divisions of living things. Some texts list as many as five major divisions.

Question 39:

What are two ways that organisms can be identified?

> a. microscope and chemical analysis
> b. plant and animal
> c. scientific study and theology
> d. hybrid crossing and cross-sectioning
> e. referral and desiccation

The following is a list of the major divisions and the organisms contained in them.

THE PROTIST KINGDOM

The living organisms in the protist kingdom are generally classified by the following:

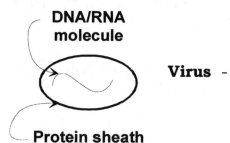

Virus - classified by the host they select. There are bacterial, plant, and animal virus. A virus invades a host and reproduces itself. Sometimes the host will die due to the infection.

Bacteria - classified according to shape. The three shapes are spherical, tubular (cylindrical) and coiled . Some examples of bacteria are diphtheria, cholera, tuberculosis, tetanus and botulism.

Most bacteria depend on living or dead organisms for their food. They are parasites (living off a host) or saprophytes (living off dead organic matter). Some bacteria have the ability to manufacture their food **photosynthetically** or **chemosynthetically** (producing chemical energy by breaking down chemicals—this chemical energy can be used to make sugars and other foods).

Examples of some common bacteria and their environments are:

Live in oxygen (aerobes):	Diphtheria Tuberculoses Cholera
Do not live in oxygen (anaerobes):	Tetanus Botulism

E. coli (Escherichia coli) is a common bacterium that lives in man and survives in both aerobic and anaerobic environments.

Reproduction in bacteria is by fission.

Protozoa - classified according to their form of locomotion.

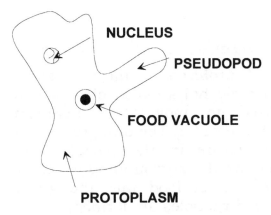

NUCLEUS

PSEUDOPOD

FOOD VACUOLE

PROTOPLASM

Amoeba (Sarcondina) - The amoeba is unique in that it has the ability to change the consistency of its protoplasm from a gel-like substance to a thick, watery solution. This solution flows from one part of the cell toward the direction the amoeba desires to move. This forms a **pseudopod**. The vacated portion of the cell follows along. This action is called protoplasmic streaming. The overall movement of the amoeba is called amoeboid movement.

Amoebas capture their food by enveloping victims with their pseudopods, forming a food vacuole.

Reproduction is by fission.

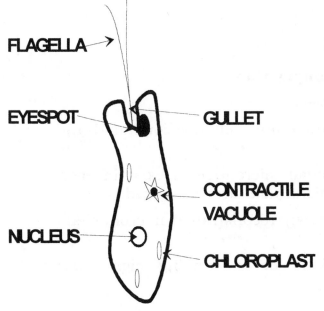

FLAGELLA

EYESPOT

GULLET

CONTRACTILE VACUOLE

NUCLEUS

CHLOROPLAST

Euglena (Mastigophora) -

These organisms have one or two long whip-like threads called **flagellum** (pl. flagella). The flagella are rotated and pull the organisms through the water. **Euglena** is a plant/animal. During the periods of sufficient light, the chloroplasts carry on photosynthesis to make food. However, at night or during periods of darkness, food is absorbed through the cell membrane and eventually digested.

Reproduction is by fission.

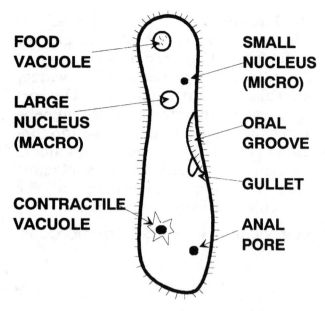

FOOD VACUOLE

LARGE NUCLEUS (MACRO)

CONTRACTILE VACUOLE

SMALL NUCLEUS (MICRO)

ORAL GROOVE

GULLET

ANAL PORE

Paramecium -

These protozoa have short protoplasmic hairs called cilia, which are used to move the organism through the water. The oral groove is a unique feature in the paramecium. Cilia agitate the surrounding water by fanning bits of food into the gullet where food vacuoles are formed.

Reproduction is by fission (mitotic division) and conjugation (two parents exchanging information, see below).

Steps of Conjugation

(1) Two cells join at the oral grooves.

(2) The micronuclei undergo numerous cell divisions and the macronuclei disappear.

(3) Eventually, a large and a small micronucleus exist in each cell. At this time, the small micronuclei are exchanged.

(4) The cells separate and the newly reconstituted micronucleus undergoes more divisions.

(5) Two further mitotic divisions occur forming eight new paramecia.

Malaria - has no form of locomotion. Depends on host for transfer from place to place. Reproduction occurs when the nucleus fragments. The fragments are surrounded by cytoplasm and are called spores. The host cell eventually ruptures, releasing the spores to infect other cells.

Question 40:

How are viruses classified?

    a. by type of host
    b. by type of locomotion
    c. by type of reproduction
    d. by size
    e. by color

Question 41:

Protozoa that move with pseudopods are called

    a. euglena
    b. paramecium
    c. stenator
    d. amoeba
    e. pseudopods

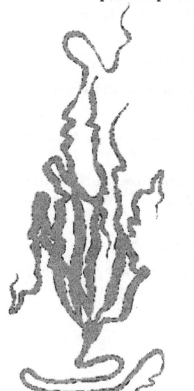

Kelp is a type of brown algae that lives in colonies.

Algae - classified according to color. Most look like plants such as seaweed and kelp. However, the blue-green algae looks like bacteria. Many algae make their food by photosynthesis.

One common use for red algae is the manufacture of **agar**. Agar is the most common medium for growing bacteria in a laboratory. First the agar is warmed up and mixed with nutrients. It is then poured into petri dishes. When it cools it forms a gel. When a scientist wants to grow a particular bacteria or mold, he pokes the sterile gel with it. This infects the gel and a whole colony of the infecting agent grows.

Blue Green Algae

CYANOPHYTA

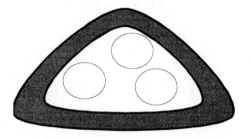

This is the type of algae often associated with scum in ponds and standing water. It is generally found in warmer areas and does not grow well in colder water.

Green Algae (Spirogyra)

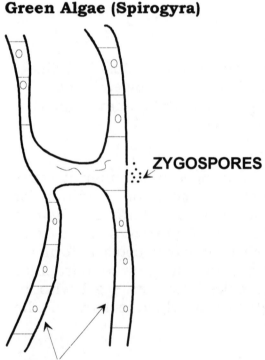

ZYGOSPORES

SPIROGYRA FILAMENTS

This is the type of plant commonly found in fish tanks. Spirogyra are different because they reproduce by both fission and conjugation.

Steps of Conjugation

1) A connecting tube forms between two spirogyra filaments.

2) The interior of one cell flows through the tube into the other cell.

3) New algae cells **(Zygospores)** are formed by the combination of two similar gametes.

4) Zygospores fall from the cell.

5) The zygospores form new spirogyra filaments if conditions are suitable.

Golden Brown Algae

**DIFFERENT SHAPES
SIZES AND DESIGNS
OF DIATOMS**

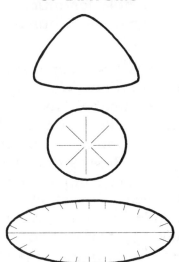

Diatoms are very abundant in the oceans. They are different from other algae because of the shell they build around themselves. This shell is made from **silica** extracted from seawater. The shells vary in size, shape, and design, but they all must be thin enough so that the algae is able to carry on photosynthesis. If they were too thick, light would not be able to penetrate the shell and photosynthesis would not occur. After the algae produces sugar through photosynthesis it quickly converts most of it to oils for storage. Many scientists believe that prehistoric diatoms, not dinosaurs, are the source of the petroleum oil we use.

Fungi - classified by shape. All fungi are parasitic or saprophytic - that is, they depend on other living or dead things for their food.

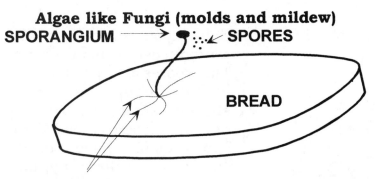

Algae like Fungi (molds and mildew)
SPORANGIUM → SPORES
BREAD
RHIZOIDS

Molds have structures that are similar to roots (**hyphae** rhizoids), but they are not true roots. Instead they penetrate the source of food and secrete enzymes. The enzymes break down the food into usable nutrients. The nutrients must be small enough to be absorbed by the hyphae.

Molds create spores for reproduction. The mold creates a shoot called a sporangium. When it matures, it releases a large amount of spores, which will form new fungi if they land in a moist environment.

71

Sac-Shaped Fungi

YEAST

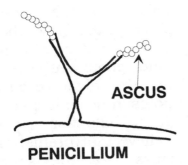

ASCUS

PENICILLIUM

These fungi also reproduce by spores. The spores are formed in an oval sac called the **ascus**. These fungi are responsible for much of the damage done to human food supplies.

Yeast is also responsible for fermentation. We use it to make alcohol (beer, wine and distilled spirits). It is also commonly used in baking. Breads that need to rise before baking use yeast. The yeast gives off carbon dioxide in the process of making energy. The carbon dioxide forms air pockets, causing the bread to rise.

Club Fungi (Mushrooms)

This type of fungi also reproduces by using spores. The spores are produced in a club shaped (mushroom shaped) structure. The top of a mushroom is called the cap. They also have root-like structures called **hyphae**. The hyphae are responsible for collecting the food.

Imperfect Fungi - Imperfect fungi reproduce asexually. They do not produce spores. Common types of imperfect fungi are ringworm and athlete's foot.

Slime Molds - Slime molds are similar to fungi, but they are mobile. They are able to move around like an amoeba. Their reproduction also involves making spores in sporangia. Another difference is the fact each of their cells contains many nuclei.

Lichens

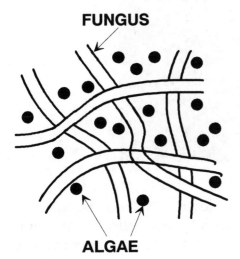

FUNGUS

ALGAE

Lichens are a unique kind of life form where algae and fungi grow together **symbiotically** (to the benefit of both). The algae provide the food and the fungi provide the water. They are able to exist in a much wider environment living together than if they lived separately.

Question 42:

How do most algae get their food?

      a. by eating animals
      b. by eating plants
      c. by eating dead material
      d. they make their own
      e. by eating fungus

Question 43:

How do lichens get their food?

      a. by eating animals
      b. by eating plants
      c. by eating dead material
      d. the algae make it
      e. the fungus makes it

THE PLANT KINGDOM

The different types of plants you should be familiar with are the ferns, the conifers, and the flowering seed plants.

Ferns - The leafy part of a fern is called a **frond**. The root is called a **rhizome**. Ferns release **spores**, which after dispersal, land on the soil, develop sperm and ova, fertilize, and become the next generation of ferns.

Let's take a closer look at the reproduction process. It occurs in two stages. As the fern matures, reproduction becomes the prime function of the plant. Blister-like **sori** form on the underside of the fronds. These sori contain the sporangium from which spores will be dispersed. The spores fall into a moist environment and germinate. This is the gametophyte stage of reproduction. The spores forms a heart-shaped (♥) **prothallus** where ova (female) and sperm (male) develop. The sperm eventually swim over to the ova and fertilization occurs. The zygote is the first part of the **sporophyte** stage of reproduction. The zygote divides and cells specialize until a new fern is formed. Just remember, ferns have a whole other stage of development that is not visible.

Conifers - These trees produce **cones**, which house the seeds. These plants stay green year-round. EXAMPLES: junipers, pines, spruces, firs, cedars and sequoias.

This type of tree is harvested extensively to make lumber products and also for making paper.

Question 44:

Which type of plant stays green all year round?

      a. flowering seed plants
      b. conifers
      c. ferns
      d. rhizomes
      e. eternals

Question 45:

What is the leafy part of a fern called?

      a. a frond
      b. a rhizome
      c. an ovum
      d. a spore
      e. a leaf

Flowering Seed Plants - This is the most varied group of plants on earth. Their seeds are equipped with a food supply, which the new plants use for food until they develop their first leaves. The fact the plants start with a supply of food allows them to get established before they have to rely on their own food production. There are two major classifications of flowering seed plants.

The first type of plant is the **monocots.** They have a single (mono) first leaf. EXAMPLES: grasses, corn, and lilies.

The veins in their leaves run in parallel along the length of the leaf.

It is important to know that this variety of plant is responsible for a large part of the human food production. All of the grains (wheat and corn) as well as rice are monocots.

Dicots: have two first leaves. EXAMPLES: sunflowers, mustard, maple trees, oak trees, and tomatoes.

An easy way to tell if a plant is a dicot is to look at the leaves. If the veins in the leaf branch (are netted), then the plant is a dicot. If they run straight without branching, the plant is a monocot.

ENLARGED

Question 46:

A plant with veins that run next to each other is most likely

    a. a fern.
    b. a conifer.
    c. a rhizome.
    d. a monocot.
    e. a dicot.

Parts of Flowering Seed Plants

Root System - The root systems of flowering seed plants are well established with tiny root hairs used to obtain water and minerals from the soil. They serve to anchor the plant in the soil and keep it from falling over. They are also important for the storage of food and water. Think about how much food a potato stores in its roots.

Question 47:

What two things does a plant get from the soil?

    a. water and oxygen
    b. water and minerals
    c. water and carbon dioxide
    d. carbon dioxide and oxygen
    e. carbon dioxide and minerals

It is important to know something about the transport system of plants. There are two types of transport systems.

Phloem - This is the part of the plant that transfers food downward from the leaf to the rest of the plant. It is formed of live cells and it is found near the outer surface of the plant. The food is transferred from cell to cell down the plant. The cells in the phloem tend to be very long because transfer occurs much quicker inside each cell than it does when it has to go between cells.

Xylem - This is the other part of the transport system in a plant. It transfers water and minerals upward in the plant. It is composed of the remains of dead cells. The ends of the cells dissolve and a hollow tube is left. For many years scientists wondered how water was able to go from the roots to the top of a tall tree. It would require an enormous amount of energy to pump the water up, and it would be impossible to suck the water up (The maximum height that a perfect vacuum can raise water is thirty-two feet). The current theory is called the Transpiration Tension Theory. Scientists believe as the leaves use water, new molecules are pulled along to replace them because of the surface tension of water. The water consists of a continuous chain of molecules from the root to the leaves.

Question 48:

What is transferred downward in a plant?

      a. water
      b. minerals
      c. glucose
      d. wastes
      e. both a and b

Question 49:

What is transferred upward in a plant?

      a. water
      b. minerals
      c. food
      d. wastes
      e. both a and b

Stem - The stem consists of bark (outer covering and protection), cambium (phloem), wood (xylem), and pith (the center most part of woody stems).

Question 50:

What is the center of the stem called?

      a. bark
      b. cambium
      c. xylem
      d. pith
      e. sap

Leaf Cross-section

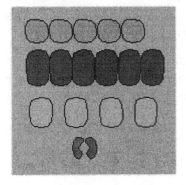

Cuticle

Palisade Layer

Spongy Layer

Stomata

Leaf - Manufactures food for the plant through the photosynthesis process. The cells responsible for photosynthesis are called **paltsadw** cells and are contained in the palisade layer. The **cuticle** is the waxy covering of the leaf. The waxy covering greatly reduces the amount of moisture lost by the leaves. Leaves have holes in them, which are used to regulate the passage of carbon dioxide (CO_2) and water (H_2O). These holes are called **stomata**. The **veins** of the leaf are the xylem and phloem.

Question 51:

If a cell needed a greater concentration of carbon dioxide, which cells would be affected?

    a. paltsadw
    b. cuticle
    c. stomata
    d. vein
    e. xylem

FLOWERING PLANT REPRODUCTION

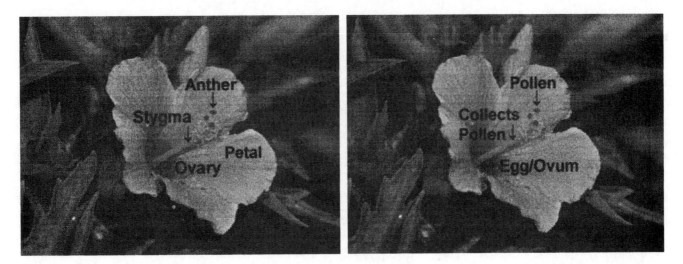

Pistil - The sticky top is called a **stygma**. The tube is called a **style**. The base of the tube is called the **ovary**. This part of the plant is "female".

Stamen - The stalk supports the **anther**, which holds the pollen. This is the "male" part of the plant. Pollen matures into sperm.

Some plants are able to self-pollinate (can fertilize themselves) while others must cross-fertilize (one plant supplies the pollen and one plant supplies the egg). Once fertilization occurs, a zygote forms and the plant puts all of its energy into seed (fruit) production.

Question 52:

What is the female part of the plant?

> a. the pistil
> b. the stamen
> c. the style
> d. the ovary
> e. the anther

THE ANIMAL KINGDOM

The animal kingdom is split into two major classifications: **Invertebrates** (without backbones) and **vertebrates** (with backbones).

INVERTEBRATES

Porifera - sponges

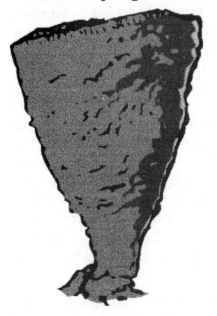

Sponges are filled with holes. Sponge cells draw water through the holes (called **incurrent pores**) and remove nutrients as well as oxygen. They then return the water to its surroundings through larger holes (called **excurrent pores**).

Cells of a sponge are arranged in two layers. On the outside, there are **epidermal** cells for protection. Inside, there are cells known as **collar cells**.

Each collar cell has a flagellum. The flagellum beat rapidly, which causes a current to flow past the collar cells. Here, food and oxygen are removed and passed along to cells called **amebocytes**, which distribute the nutrients to other cells. The water then continues to flow out the excurrent pores.

Sponges have **spicules**. They make up the skeleton of the sponge. They are what remains after a sponge dies.

Sponges are able to reproduce in three ways. The first is by **budding**, reproduction by breaking off a piece of the original to form two individuals. The second is **regeneration**, or replacing tissue that is lost. They may also reproduce sexually by releasing sperm in the water. The sperm swim around and enter other sponges and fertilize the ova. The zygote eventually develops into a "free swimmer" that will eventually settle to the ocean floor and become a sponge. This allows sponges to spread out to new areas.

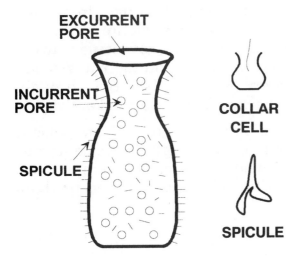

EXCURRENT PORE

INCURRENT PORE

SPICULE

COLLAR CELL

SPICULE

Coelenterates - Jellyfish and Corals

Coelenterates exist in various body shapes during their life span. Some begin life as free-swimming **planula**. Planula are the larva of coelenterates. Some planula develop into a polyp-shaped **hydra** or into a **medusa** (a type of mature coelenterate that lacks means of self-propulsion). Both adult forms of coelenterates are able to reproduce. Other coelenterates have planula and some form alternate generations. Planula develop into polyps, which then produce more planula. They then develop into medusas, which produce more planula in order to complete the cycle.

Coelenterates contain two layers of cells separated by a jelly-like substance called **mesoglea**. The stinging cells located in the **tentacles** are called **nematocysts**. They contain paralyzing poisons. Food is obtained when several stings from the nematocysts paralyze a victim. Slowly, food is drawn into a gastrovascular cavity (in polyps) where it is thoroughly digested. Reproduction may occur through budding, regeneration, or sexual activity.

There are some fish, clown fish for example, which are immune to the poisons in the nematocysts. They live in a symbiotic relationship with the coelenterate. The clown fish swim among the tentacles and when another fish comes to eat it, the predator fish gets stung. The clown fish feeds off the poisoned predator fish.

Worms - There are three distinct types of worms. They show increasing specialization and complexity.

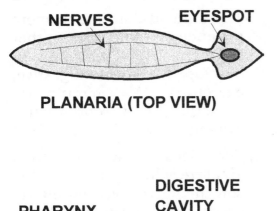

PLANARIA (TOP VIEW)

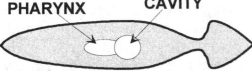

PLANARIA (BOTTOM VIEW)

PLANARIA (SIDE VIEW)

Platyhelminthes - Flatworms (Planaria)

These are the least complex of the worms, but they are much more complex than the organisms that have come before them. They are the first organisms to have three layers of cells: **ectoderm**, **mesoderm** and **endoderm**.

The planarian (flatworm) is a free-swimmer which sucks its food through the pharynx tube.

It may reproduce by asexual (fission) or sexual means. The animal contains both male and female sex organs (**hermaphroditic**) so cross-fertilization is easily accomplished.

The planarian has marvelous powers of regeneration. When it is damaged it can regenerate up to half its body.

Nematodes - Round Worms (Hookworm, Trichina)

These worms are a little more complex than flatworms. The biggest difference is their body shape. They are round instead of being flat, as their names suggest. They are the first animals that have a digestive system with two openings, one for the mouth and one for the anus. Flat worms only have one opening which serves both functions.

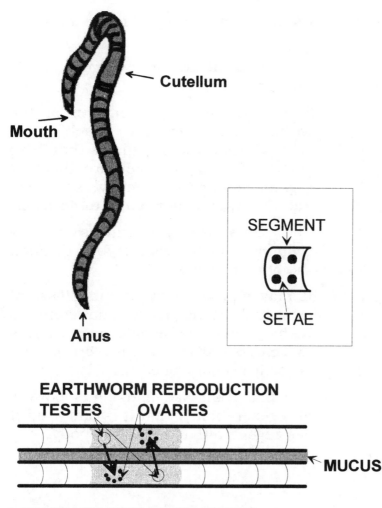

Annelids - Segmented Worms (Earth Worm)

The annelids are much more advanced than the other worms. They have well-developed nervous, digestive, excretory, circulatory, muscular, and reproductive systems.

The digestive system has a **gizzard**, which grinds up the soil the earthworm swallows. Digestive enzymes attack the finely ground up soil, extracting nutrients. No longer is the food simply passing through the body.

After the food is digested, the circulatory system takes over. Annelids have multiple hearts and a circulatory system that distributes food over the length of the body and eliminates waste products.

Earthworms have "grippers" called **setae** on their underside. This allows them to anchor themselves into the ground.

Reproduction is sexual; one worm supplies the sperm and the other the egg. However, one interesting feature is earthworms are hermaphroditic. That is, they have both male and female sex organs. They cannot self fertilize; they must find another worm with which to mate. The two worms are able to simultaneously fertilize each other.

Mollusks - Shellfish (Clams, Oysters, Scallops)

Mollusks have soft bodies and an outer layer of cells called the mantle. There are three major categories of mollusks.

Gastropods - "Stomach Foot" - (Snails)

Most of these mollusks have a shell. (Slugs have none at all!). The shell is in a spiral shape and the animal has its vital organs protected inside the shell.

Gastropods have eyestalks, which are flexible, so that they can move their eyes to look in different directions.

Gastropods "breathe" through the exchange of gases in the **mantle** cavity. Oxygen dissolves in and carbon dioxide dissolves out.

Gastropods are capable of movement, although it tends to be slow. They secrete slime and contract the foot muscle to pull themselves forward. When they are scared, they pull the foot and the rest of their body back into their shell for protection.

Eating is accomplished by scraping its **radula** (made of chitin) against leaves and digesting nutrients in a fully developed digestive system. Some Gastropods eat other mollusks.

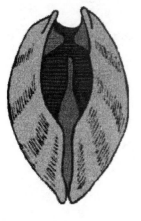

Pelecypods – "Two Shells" - (Clams, Oysters]

As the name suggests these mollusks have two shells. Some bi-valves have symmetrical shells (both are the same); others have one shell that is larger than the other shell. They can vary greatly in size, ranging from under half an inch to the enormity of giant clams, which weigh hundreds of pounds.

Bi-valves contain two **siphon tubes**. One is used to bring nutrients in, logically called the incurrent tube, and one sends waste out, called the excurrent tube. The flow of water through the tube sucks in food. The food is captured by the mouth and then sent to the digestive system. The nutrients from digestion are moved around with a circulatory system while the waste products of digestion are excreted through the anus. The circulating water also flows over the gills. Here, oxygen is removed from the water and carbon dioxide waste is put into the water to be sent out the excurrent tube.

Cephalopods - "Head Foot" - Octopus, Squid

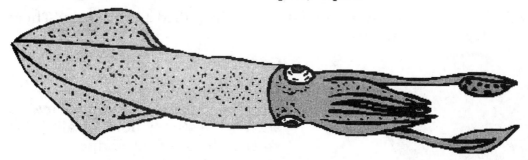

Squid are free-swimmers that use "jet-propulsion" to move around their marine environments. Their tentacles have sucker discs with which squid are able to grab food. Evidence of epic battles between giant squid and sperm whales have been found on the whales where some scars from the sucker disc have been over a foot in diameter. Squid must rely on their speed for protection.

While octopi can move by jet propulsion, their more common mode of movement is by walking along the bottom on their tentacles. Often they are found with their bodies in rock crevasses for protection. When caught in the open, an octopus will excrete a cloud of ink, which hides it while it escapes.

The eyes of cephalopods are more advanced than other mollusks. Their digestive and circulatory systems are also more advanced than the other mollusks.

Arthropods - "Jointed Foot"

This is the largest and most varied phylum of animals on the earth. All arthropods have the following common characteristics:

1. A hard outer body shell called an **exoskeleton**. This exoskeleton is made of a substance called **chitin**.

2. These are the first animals to have **jointed appendages**.

3. There are four classifications of arthropods:

    a.   Myriapoda

    b.   Insects

    c.   Crustaceans

    d.   Arachnids

Myriapoda – "Many Feet" - Centipedes, Millipedes

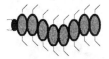

**CENTIPEDE
TWO LEGS PER
SEGMENT**

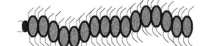

**MILLIPEDE
FOUR LEGS PER
SEGMENT**

There are two major branches of the Myriapoda, centipedes and millipedes. Centipedes have one set of legs per segment, and millipedes have two sets of legs per segment. Centipedes are carnivorous and contain poison claws, which enable them to capture food. Millipedes are mainly herbivorous.

Insects - Mosquitoes, Flies

Insects are responsible for a lot of crop damage, but there are also some insects that are beneficial to humans. Many farmers release Ladybugs to eat aphids and other insects, which would harm crops.

There are some distinct characteristics that separate insects from other arthropods. Insects have three major body parts: head, thorax, and abdomen. They also have six legs; breathing tubes called **spiracles** located in the abdomen; one pair of antenna; and two pairs of wings.

The insect life cycle, coupled with wings and special feeding parts, has allowed insects to inhabit all but the coldest portions of the earth's surface. They are often the first animals to move into a new area such as a newly formed volcanic island.

Most insects undergo a **metamorphosis** (change in body shape) from egg through to adult form.

1. Egg
2. Larva (worm-like—caterpillar, etc.)
3. Pupa (change from worm-like form to adult form)
4. Adult

Not all insects go through all phases.

Insect eggs that hatch into small adult forms, called **nymphs**, are **bugs**. A good example of a bug is a grasshopper.

Crustaceans - Lobsters, Crabs, and Shrimp

The big difference between crustaceans and insects is that the head and the thorax are joined together in one part the cephalothorax. The abdomen remains the same as in insects. Crustaceans are the largest of the arthropods with lobsters growing to 40 pounds or more. Crustaceans have gills. Even the crabs that live on land must keep their gills moist in order to transfer oxygen and carbon dioxide.

Arachnids - Spiders, Scorpions, Horse-Shoe Crabs, Mites and Ticks

Arachnids have some features that are similar to crustaceans. Arachnids also have only two body parts: the cephalothorax and abdomen.

Spiders have **book lungs** for the transfer of oxygen and carbon dioxide. They also have exactly eight legs: no more, no less.

Another distinguishing feature is they have no antennae.

Echinoderms - Starfish

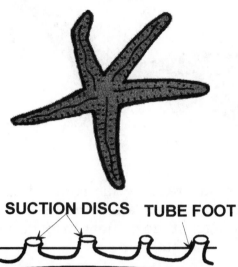

SUCTION DISCS   TUBE FOOT

WATER VASCULAR SYSTEM

Starfish have hard, spiny bodies. The underside of each arm has two rows of **tube feet.** These feet are manipulated by the **water vascular system**.

When water is pumped into this system the **suction discs** puff out and release their grip on the surface. When the water is moved back out, the tube feet can attach to something using suction. By rhythmically moving the water through the system, alternating the attachment/release of the discs, the starfish is able to move through the water.

The tube feet are also used to capture oysters or clams for food. The starfish attaches to the clam and starts to pull it apart. At first the clam is strong enough to keep its shell closed. Over time however, the clam tires and its shell opens a little. The starfish then slips its stomach inside the shells and digests the contents.

Starfish reproduce sexually. Male and female starfish release sperm and ova into the water. Fertilization occurs and produces ciliated larva (similar to that of paramecium). The larva are able to move around and often drift long distances with the ocean currents. Eventually they settle to the bottom and start their lives as adults.

Adult echinoderms exhibit radial symmetry. Most other adult animals are bisymmetrical (there is only one way to cut them down the middle and end up with both halves being alike.) Animals with radial symmetry can be cut in more than one way to get identical halves.

Chordates

Chordates are the most advanced form of animals. Chordates contain three prominent features at some time in their lives. Some animals, like humans, only exhibit some of these features while they are developing before they are born. They all have:

1. **Notochord** - a hard, internal, supporting rod of connective tissue.

2. **Paired Gill Slits** - become gills in fish and amphibians; disappear in other chordates (including man).

3. **A Dorsal Nerve Cord** - spinal cord.

The simplest of the chordates are the sharks and rays. While some sharks are very efficient hunters, they are not as evolutionarily advanced as fish. A shark is not a fish. Its skeleton is made of cartilage; it does not have bones. Its notochord also is not enclosed in a backbone for protection.

VERTEBRATES

The rest of the animals listed will all be vertebrates. Vertebrates have their notochord surrounded by a backbone for protection. They also have bones in their skeleton

Pisces - Fish

These are the simplest of the remaining animals. Their heart only has two chambers, so the blood coming from the gills mixes with the blood coming from the body in the heart.

Fish have scales on the outside of their bodies and rely on gills for breathing.

Many fish have a swim bladder. This cavity can be filled with gas, which allows the fish to adjust its buoyancy to different depths in the water.

Fish reproduce sexually. Often the eggs and the sperm are simply released in the water to float around freely. It is left to chance that they will meet up and fertilize each other.

The fins on a fish serve as stabilizers. They keep the fish from spinning out of control in the water.

Amphibians - Frogs

Adult amphibians tend to live in moist environments. They have lungs, but need to keep their skin moist. Many amphibians go through a juvenile stage and then they go though a metamorphosis to the adult phase. A frog is a good example. When it first hatches out of the egg it still has the yolk sac attached. It has more structure that is similar to fish than it does to adult amphibians. It has a long tail for swimming, gills and a two-chambered heart. It cannot survive on land and it has no legs. After the tadpole lives like this for awhile and stores up enough energy, it will undergo metamorphosis. It uses the energy stored in its tail to change into the adult form. The first thing to appear is the back leg. Slowly it will also develop smaller front legs, lungs and a three-chamber heart. It will now be able to survive on land as long as it keeps its skin moist. It will also be at home in the water as long as it is able to come up for air and breath. It will no longer be able to survive exclusively in the water because it has lost its gills.

In areas where it gets colder in the winter, the frogs will hibernate. They will go down into the mud at the bottom of a pond to hibernate. All of its processes will slow way down. It lives off the food it stored over the summer. Because its metabolism (how fast it is using energy) has slowed way down frogs do not have to breathe air when they hibernate. They are able to transfer enough oxygen and carbon dioxide through their skin.

Amphibians reproduce sexually. The female lays its eggs in water and the male fertilizes them. They can often be found in clumps in ponds.

Reptiles - Turtles, Alligators, and Snakes

Reptiles are even more advanced than amphibians. They have made the transition to true land animals, although some have returned to the water to live. A big advance for reptiles is that they have a four-chambered heart. A four chamber heart allows reptiles to keep the oxygenated blood from their lungs separate from the blood laden with carbon dioxide that is returning from the rest of the body. Reptiles also have four legs, although as snakes have developed their legs have shrunk back into their body and all that is left are the bones. Reptiles are **cold-blooded**; they cannot control their body temperature. This means that they are only able to move very slowly when they are cold. To counteract this you will often see reptiles sunning themselves (laying on a rock in the sun). This raises their body temperature. Reptiles have thick, scaly skin.

Reptiles reproduce sexually. Fertilization occurs internally as the male places the sperm in the female. Most of the time the female then lays eggs, however there are some reptiles whose young are born alive directly from the mother.

Aves - Birds

Birds are the next step up evolutionarily from reptiles. They are the first animals that are **warm-blooded**; they are able to control their body temperature. Their feathers provide them with insulation. When it gets cold, they fluff out their feathers to trap more of the warm air close to their bodies.

Birds, like reptiles, have a four-chambered heart. An interesting feature birds have is their wings. Although not all wings are used for flight, all birds have them. They also have hollow bones. This makes their bones much lighter than other animals. If they had solid bones like other animals they would be too heavy to fly.

Bird reproduction is similar to reptiles. They reproduce sexually and fertilization takes place internally. The eggs are then often laid in a nest. The eggs are covered with a calcium-based shell. In the twentieth century many large birds were having problems. Humans were spraying for insects with DDT. The birds were secreting the DDT in their eggshells, which made them very fragile. When the birds sat on the nest to keep the eggs warm, the shells cracked and the eggs died as a result.

Most birds show many adaptations from their basic structure depending on where they live and what they feed on. A bird's beak will be different depending on if they feed on insects or seeds. Their feet also show adaptations. The long talons of eagles and owls are designed for grabbing their food.

Mammals - Bats, Whales, Porpoises, Platypus, Kangaroos, and Humans

Mammals possess a body covering of hair, four-chambered hearts, a diaphragm to separate the chest cavity from the digestive cavity, a highly developed brain, and **mammary glands** which provide nourishment (milk) for the newborn.

Monotremes – Platypus and Echidna

These are the least advanced of the mammals. They lay eggs like birds. Their mammary glands are not very developed. The milk just sort of oozes out when the young feed. They are found only in Australia and its surrounding islands. Competition with more advances mammals eliminated monotremes from the rest of the world.

Marsupials – Kangaroos and Opossum

While there are a few marsupials found throughout the world, the vast majority are found in Australia and its surrounding islands. Here marsupials have adapted to all of the niches filled by more advanced mammals in the rest of the world.

Marsupial babies are born alive, but very young. They then crawl up to the mother's pouch. Here they feed on a mammary gland and continue to develop. They will spend the first weeks of their lives exclusively in their mother's pouch. Even after they are finally big enough to emerge on their own, they will still return to their mother's pouch for protection.

Placentals - Mice, Bears and Humans

These are the most advanced of the mammals. Their young are generally born fully developed, although some are not fully capable of surviving on their own at the time of birth. The young tend to spend a long time in the **womb** (the place inside the mother where the babies develop) before they are born.

Placentals are the most advanced animals and have a higher survival rate than many other animals because their young have such a long period of development inside the mother. For this reason, the birth rate of placentals tends to be lower than many other animals. Some placentals only give birth every couple of years.

Question 53:

Which of the following are vertebrates?

    a. mollusks
    b. aves
    c. echinoderms
    d. a and b
    e. all of the above

Classification Rankings

All of the different classifications mentioned can be broken down into different groups. In order of increasing specialization, they are: Kingdom, Phylum, Class, Order, Family, Genus, and Species. Kingdoms are the broadest category (Plant and Animal) and Species are the smallest. All members of a given species are able to interbreed and produce **viable** offspring (young that survive and can reproduce).

If you are asked which group has more members, the order that an organism is in always has more members than its Family, Genus, or Species.

HUMAN SYSTEMS

The Male Human Reproductive System

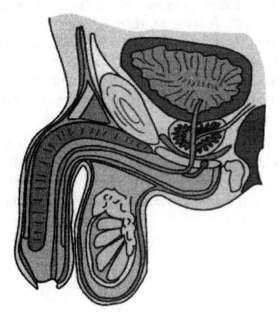

Sex cells are produced in the **testes**. They grow tails (**flagella**) and are called **sperm** or **spermatozoa**. The tail's function is to provide locomotion in the fluid called **semen**.

Testes are held in a sac called the **scrotum**, which is located below, and outside the abdomen. The location is critical because sperm production and storage is best at lower than body temperature (98.6 F).

The duct system from the testes through the penis involves the **epididymis**, the **vas deferens**, the **ejaculatory duct**, and the **urethra**. The sperm travel through this passageway of ducts to exit the male body. The **prostate** and **seminal vesicles** provide semen and chemicals to activate the sperm.

Question 54:

Why must sperm cells develop outside the body?

        a. To keep them warm.
        b. To keep them cool.
        c. To keep them wet.
        d. To keep them dry.
        e. None of the above.

Question 55:

How does sperm move?

        a. with pseudopods
        b. with cilia
        c. with flagella
        d. with legs
        e. with scrotum

The Female Human Reproductive System

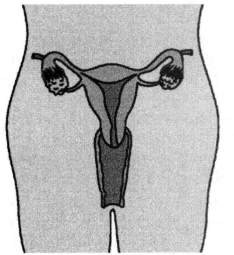

Female sex cells are produced in the **ovaries**. Female sex cells have no means of locomotion. The cells are called **ova** (pl.).

The ovaries are located in the lower abdominal cavity and produce the hormones **estrogen** and **progesterone**.

An egg (ovum) is passed from the ovary to the **fallopian tube** (oviduct) where it is moved along by the ciliated (small hair-like projections that beat back and forth) lining of the tube.

If fertilization has occurred in the fallopian tubes, the **zygote** attaches to the walls of the uterus and grows into a **fetus**.

If fertilization has not occurred, the female experiences **menstruation**, and the inner layers of the uterus are discharged.

The **vagina** is the cavity that receives semen from the male. It also discharges menstrual flow and is the passageway for birth.

External features include the **vulva** (labia majora, labia minora), which encloses the vagina, and the **clitoris**, which is a small erectile, sensitive tissue.

Question 56:

Which of the following is true if menstruation occurs?

        a. implantation occurred
        b. the egg was not fertilized
        c. menopause
        d. a zygote is formed
        e. birth

Question 57:

Where does the egg implant if it is fertilized?

        a. in the uterus
        b. in the fallopian tubes
        c. in the vulva
        d. in the vagina
        e. in the ovary

The Circulatory System

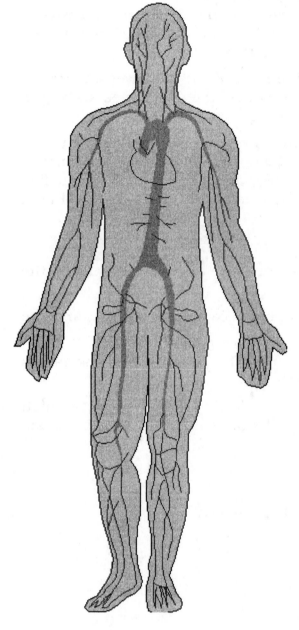

The function of the circulatory system is to provide the flow of materials through the body. Circulation provides for the exchange of gases (O_2-CO_2), removal of wastes, regulation of body temperature, nutrition, H_2O regulation, proper pH maintenance, and immune protection (white blood cells and antibodies).

Blood is divided into two parts, the plasma (fluid) and the cells. The **plasma** contains water, carbonates, chlorides, phosphates, urea, hormones, vitamins, digested food, albumin, globulin, fibrinogen, and prothrombin.

There are three different types of blood cells.

1) **Red Blood Corpuscles (erythrocytes)-**

The main function of this type of cell is to transport oxygen. The cells contain a lot of hemoglobin. It can readily combine with and disassociate from oxygen. It all depends on the surrounding concentrations of oxygen. When there is a low surrounding concentration it releases oxygen. When the surrounding concentration is high, it accepts oxygen. Therefore, red blood cells pick up oxygen in the lungs and distribute it to the rest of the body. Mature red blood cells have no nucleus (so they can carry more hemoglobin) and are shaped like a donut without the hole (to increase their surface area). More surface area means quicker transport from the cell to the outside.

White blood corpuscles (leukocytes) - White blood cells are the second type of blood cell. They contain no hemoglobin. Their job is not the transfer of oxygen. Their job is in fighting infections. They are **amoeboid** in shape and method of locomotion. They roam around the blood stream looking for invaders (bacteria and viruses). When they find them they eat them. They work closely with antibodies. When the body gets an infection it produces antibodies. Antibodies cause viruses and bacteria to group together. This makes it easier for the white blood cell to catch and eat the infecting organism.

White blood cells are not confined to the blood stream (arteries, veins, and capillaries). They are able to squeeze out through the walls and go out among the cells to where the infection is.

Platelets (thrombocytes) - Platelets are important in **clotting**. If we did not have platelets, the simplest cut would cause us to bleed to death. When we are cut platelets rupture at the site. This starts the clotting process. A net is made of thrombin and fibrinogen fiber, which traps red blood cells. These cells then dry into a clot.

All of the blood cells are produced in the **bone marrow** with the exception of leukocytes, which may also form in the **spleen**.

Question 58:

Which type of blood cells protects you from bleeding to death when you cut your finger?

     a. red
     b. white
     c. platelets
     d. protectors
     e. none of the above

Vessels carrying blood include:

1) from heart to body - **aorta, arteries, arterioles** (the smallest arteries)

2) **capillaries** - small blood vessels connecting arterioles to veinules

3) from body to heart - **veinules** (the smallest veins), **veins, vena cava**

Capillaries - the site where nutrients are exchanged.

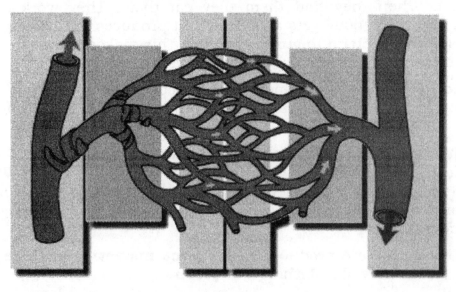

Arteries have thick walls and are under pressure from the heart pushing blood through the body. The capillaries connect the arterial system to the **venous** system. Veins have thin walls and low pressure. The blood is moved along veins when muscles contract. Valves prevent the blood from flowing backward.

The **heart** is a muscular pump, which causes the blood to circulate through the body. It is **cone-shaped** and has walls, which have three layers of muscle tissue: **endocardium, myocardium**, and **pericardium**.

Blood flows back to the heart from all parts of the body. The **venae cavae** empty into the **right atrium** of the heart. Blood flows to the **right ventricle** and then out to the lungs through the **pulmonary arteries** (the only non-oxygenated arterial blood in the body). After the exchange of gases in the lungs, blood returns through the **pulmonary vein** (oxygenated blood) to the **left atrium**, then to the **left ventricle** and out to the body through the **aorta**.

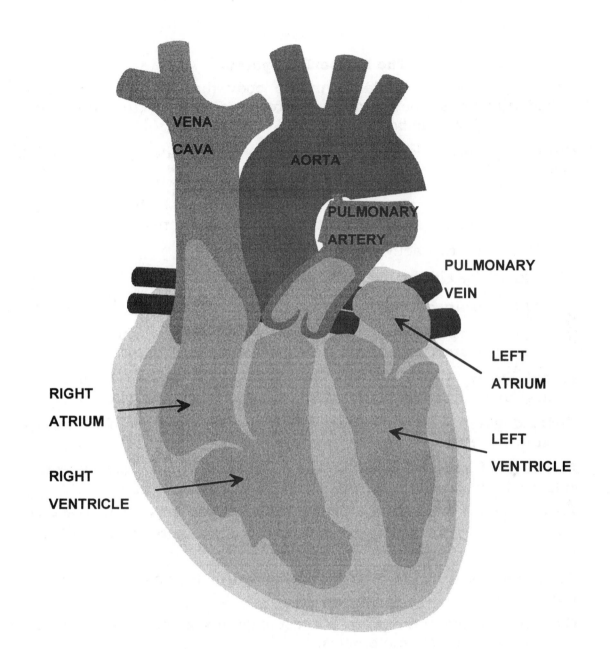

Question 59:

Which of the following is not true?

- a. The left ventricle pumps blood to the body.
- b. The veins only carry deoxygenated blood.
- c. Most of the transfer occurs in the capillaries.
- d. Valves keep the blood from flowing backwards.
- e. The pulmonary artery carries blood with a lower oxygen content than the aorta.

The Endocrine System

This system is made up of glands in the body that secrete **hormones** to regulate body functions. The hormones are released into the circulatory system and carried throughout the body.

1. **Pituitary gland** - located in the brain. It regulates or activates the following:

    (a) growth

    (b) cortex (the outer portion of the brain) activity

    (c) thyroid activity

    (d) H_2O in the blood

    (e) ovary or sperm development

    (f) corpus luteum (a ductless gland in the ovary) and testosterone

2. **Pineal gland** - located in the brain. It regulates ovaries and is the "biological clock" of the body.

3. **Thyroid gland** - located in the neck. It regulates metabolic rate and calcium concentration in the blood.

4. **Parathyroid gland** - located on the backside of the thyroid. It regulates calcium in the blood.

5. **Adrenal gland** - located on the top of the kidneys. It regulates potassium and sodium in the blood, heartbeat, blood pressure, and blood sugar level.

6. **Pancreas** - located below the stomach. It regulates passage of sugar into the cells.

7. **Ovaries** - located in the pelvis of the female. It regulates the development of sex organs and characteristics.

8. **Testes** - located below the pelvis in the male.

9. **Thymus** - located in chest. Its function is unknown.

Question 60:

Which does a gland in the brain not control?

    a. Sugar
    b. Growth
    c. Ovaries
    d. Thyroid
    e. Water in the blood

e. Water in the blood

This system consists of the **brain**, **spinal cord** and **nerves**. The function of this system is to regulate and coordinate body activities.

There are two main divisions in this system:

1.   **central nervous system** - which causes voluntary movement.

2.   **autonomic nervous system** - which regulates heartbeat, glands, and smooth muscles.

When the brain has a message to send to the rest of the body, it travels along the nerves. The junction between two nerves is called a synapse. The axon of one nerve meets the dendrite of the other. Chemicals pass across the space in between to tell the second nerve to carry on the signal.

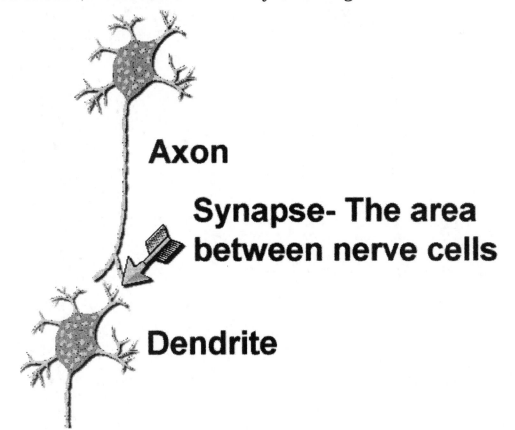

Axon

Synapse- The area between nerve cells

Dendrite

Some nerves are able to regenerate, but the nerves of the brain and spinal cord cannot.

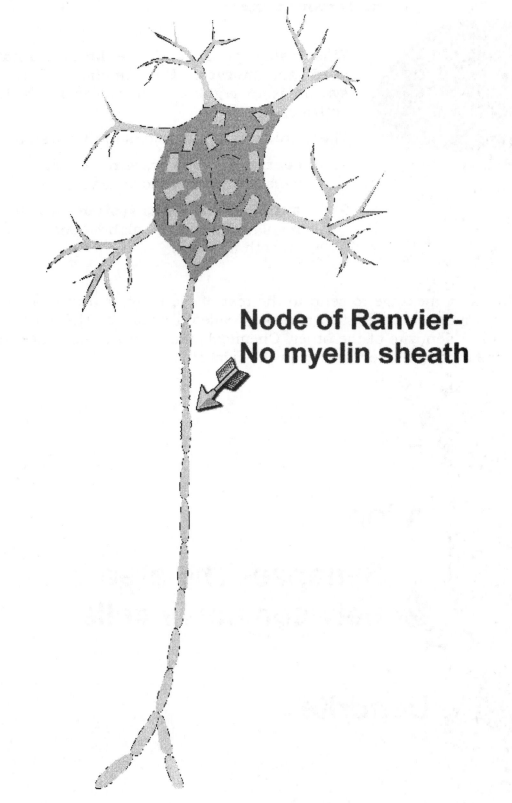

Node of Ranvier- No myelin sheath

Most nerves are covered with a **myelin** sheath. It greatly speeds up the movement of the signal through the nerve. The signal jumps from one **node of Ranvier** (site of no myelin) to the next node of Ranvier

Question 61:

Which is not part of the nervous system?

    a. Glands
    b. Brain
    c. Spinal cord
    d. Nerves
    e. None of the above

The Excretory System

Kidney

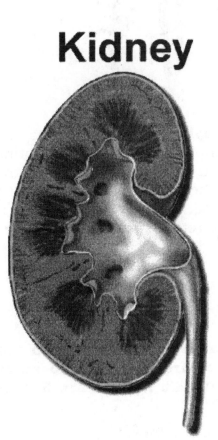

This system eliminates toxic or excessive by-products of metabolism.

In humans, excess CO_2 that builds up from the metabolism of food is removed from the body by the lungs. The **kidneys** are the primary excretory organs for other products of metabolism. The kidneys remove the end products of digestion, some excess H_2O, vitamins, hormones and enzymes. They also convert **urea** to **urine** in order to discharge waste from the body.

The kidneys perform a very important function: when the kidneys stop working the blood quickly becomes toxic. People with this problem must go in for dialysis. There, a machine takes the toxins out of the blood.

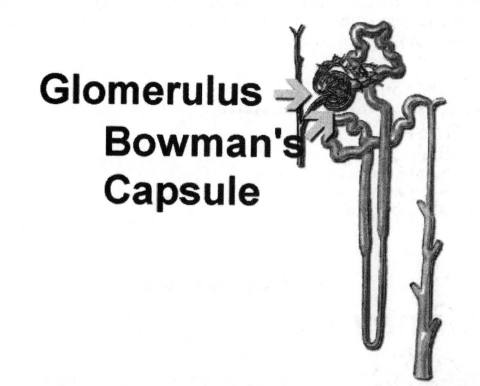

Glomerulus
Bowman's
Capsule

The kidneys are composed of many individual **nephrons**. Blood enters the nephron at the **glomerulus**. The glomerulus is composed of a ball of twisting capillaries. The blood enters under pressure, which forces fluid out into the **Bowman's capsule**. From the Bowman's capsule, the fluid flows down a long loop (the **Loop of Henle**) and the water is reabsorbed. The fluid left is urea. It is collected and stored in the bladder until it is excreted through the urinary tract.

Question 62:

Kidneys are one of the most common organs to be transplanted and the waiting list is long. Many people wait for years before they are chosen as recipients. What is one of the main reasons people can wait on the list for a new kidney longer than for some other organs?

    a. There is no need to rush out and get a new kidney.
    b. Kidney failure is not life threatening
    c. People in kidney failure can be kept alive by dialysis while failure of other organs will kill them.
    d. There are more hearts available for transplant than kidneys.
    e. Kidney transplant is much more complicated than other types of transplants.

The Muscular System

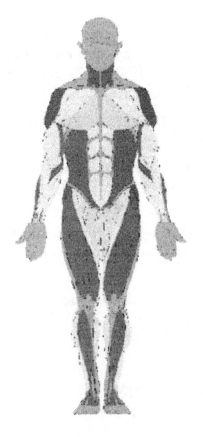

This is the system that causes movement. This system contains **voluntary** and **involuntary** muscles.

Voluntary muscles are **striated** and are used for fast contraction. They are the ones that are connected to the bones and cause them to move.

Involuntary muscles include **smooth** and **cardiac** muscle. Smooth muscle contraction is slow. A good example is the muscles that line the digestive tract. You have no conscious control over these muscles. They move the food along in a slow deliberate manner.

Cardiac muscle contracts in a regular fashion and does not tire easily. It is the only type of muscle that can contract on its own without any stimulus from nerves. If living cardiac cells are put in a beaker with a nutrient broth they will continue to contract on their own.

Question 63:

What are the two types of muscle?

    a. Voluntary and striated
    b. Involuntary and voluntary
    c. Involuntary and smooth
    d. Involuntary and cardiac
    e. Smooth and cardiac

The Respiratory System

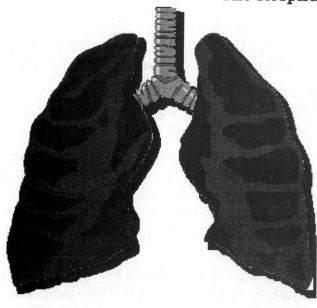

This is the process by which cells receive O_2 and give off CO_2. The steps involved in respiration are:

1. **Breathing** – the movement of air in and out of the lungs.

2. **External Respiration** – the exchange of gases (O_2 - CO_2) in the lungs.

3. **Transpiration** – the movement of O_2 - CO_2 by blood to and from cells.

4. **Internal Respiration** – the exchange of gases between the cells and the blood.

Oxygen is used with nutrients in the cell to produce usable energy.

Alveoli

Let's take a closer look at an **alveoli**. Alveoli are the functional unit of the lungs. They look like a tiny balloon. When we breathe in air inflates these little sacs. Many capillaries surround the sacs. These capillaries are carrying deoxygenated blood from the pulmonary artery. Carbon dioxide dissolves out of the blood and into the air sac. Oxygen from the sac dissolves into the blood creating oxygen rich blood. The oxygen rich blood then flows back down the pulmonary vein to the left atrium of the heart. When the carbon dioxide level in the sac reaches a certain point, nerves send a message to the brain and we exhale.

Question 64:

Which of the following is true?

    a. Oxygen is in a higher concentration in the blood than in the alveoli.
    b. CO_2 concentration is higher in the alveoli than in the blood.
    c. The oxygen level in the alveoli makes you breath.
    d. Oxygen rich blood flows in the pulmonary artery.
    e. CO_2 dissolves from the blood to the alveoli.

The Integumentary System

The **skin** provides an outer protective covering for the body. It keeps viruses and bacteria from infecting the body. It also protects us from the harmful rays of the sun. When we are exposed to a lot of sunlight our skin produces extra melanin and it gets darker.

The Digestive System

Digestive Tract

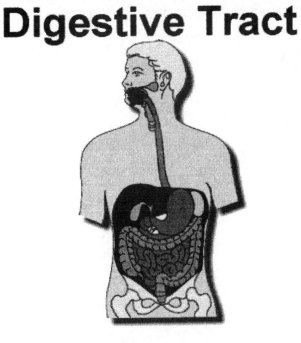

This is the process that changes large organic molecules into smaller organic molecules that can pass through cell membranes. **Enzymes** are responsible for this process.

1. **Mouth** - contains **teeth** and the **tongue**. Secretes **mucus** and **saliva** and begins the breakdown of food by enzymes.

2. **Esophagus** - moves food to the stomach area.

3. **Stomach** - churns food and mixes in **gastric juices** and **hydrochloric acid** to further break down food. Here the digestion of fats begins.

4. **Small Intestine**—more secretions are added here. They include **pancreatic juice**, **bile**, and **intestinal juices**. Food is broken into usable molecules and taken away by the blood. The small intestines contain many **villi** (small projections) which increase the surface area and allow for greater absorption.

5. **Large Intestine** (colon)—undigested food and unusable solids (feces) are compacted and water is removed. The excess waste is then removed from the body through the anus.

Question 65:

Which of the following is the path food takes through the digestive system?

      a. mouth, esophagus, stomach, small intestine, large intestine
      b. mouth, esophagus, stomach, large intestine, small intestine
      c. mouth, stomach, esophagus, small intestine, large intestine
      d. mouth, stomach, esophagus, large intestine, small intestine
      e. stomach, mouth, esophagus, large intestine, small intestine

This completes our biology review. Before you take the chapter test, be sure to check your answers to the questions appearing throughout the chapter.

ANSWERS – BIOLOGY CHAPTER QUESTIONS

If you have answered a question incorrectly, refer to the page where the question appears and review the information given there.

1. e. The mitochondria organelles are where energy production takes place.

2. b. The main function of the cell membrane is to control what enters and leaves the cell. It is the cell's first line of defense against disease.

3. d. The cell wall is non-living (dead) and made of cellulose. The first three answers can be ruled out because they say it is living. The last answer indicates that it has chlorophyll.

4. a. Plant cells lack mobility and need the protection they get from their cell walls. The other answers are functions carried out by the living parts of the plant.

5. c. Water and carbon dioxide combine to form sugar. All of the incorrect answers contain at least one of the products.

6. d. Since photosynthesis is the process that plants use to make food, they would eventually die if they could no longer do it. Animals either feed on plants or on animals that eat plants. If all plants die, animals would also eventually die.

7. e. Some plants live off other plants, and some live off of dead plants and animals. The other answers are all false. Animals and fish do not make their own food. Not every plant can make its own food, and all organisms are able to change molecules.

8. a. DNA stands for deoxyribonucleic acid.

9. e. The first four answers are all true. The fifth answer says that chromosomes grow continuously. This is not true. Chromosomes do not get longer all the time. When a cell divides it makes new chromosomes of exactly the same size.

10. b. The rungs are made of nucleic bases. The outside is made of sugars and phosphates. That means that they are not in the rungs. The paragraph says nothing about whether the rungs shift or are broken.

11. c. Genes are joined together to make chromosomes. This is the best answer. A chromosome is a type of molecule, but c is a better answer than a. Genes are not found outside the cell and do not float freely in the cytoplasm.

12. d. Adenine pairs with thymine. Cytosine and guanine would be paired together.

13. b. RNA is able to leave the nucleus. The other answers are all wrong. DNA is able to duplicate itself and form a double helix. It also stores the genes. Neither DNA nor RNA divides the cell.

14. b. Since the DNA starts out in the nucleus and cannot leave RNA must be produced in the nucleus. All of the other choices are outside the nucleus.

15. a. In an RNA strand, thymine is replaced by uracil. Adenine bonds with thymine in DNA, so it would have a different base in RNA. Cytosine and guanine always pair together in both DNA and RNA

16. c. M-RNA ends in the area of a ribosome where a mutual attraction occurs.

17. e. DNA continues to make M-RNA until it is stopped by a chemical stimulus.

18. d. There are twenty amino acids.

19. c. Transfer RNAs have amino acids attached to them. The only other type of RNA that we have talked about is messenger RNA. It is produced in the nucleus and has no amino acids.

20. c. This is a tricky question. The first answer appears correct. "T-RNA can attach to any amino acid." The key word to watch out for is "any." Each T-RNA attaches only to a specific amino acid. The same is true for answer b. T-RNA can only attach to a specific point on the messenger RNA. The last answer is wrong because it is the amino acids that form the protein not the transfer RNA.

21. a. Energy production occurs mainly in the mitochondria.

22. b. Oxidation is the opposite of photosynthesis because the products and the reactants switch. Energy is given off instead of being stored.

23. b. Looking at a formula, the reactants would be on the left-hand side of the arrow. The chemicals on the left-hand side of the formula in question are water and fat.

24. c. This is the state of the cell when it goes from step four to step one. In this phase the cell stops converting ADP to ATP.

25. a. The key words here is "small amounts" of energy. If the cell needed large amounts of energy, it would start breaking down food. When the cell only needs small amounts of energy, it converts ATP to ADP.

26. d. It is important for cells to get the same material because they would probably die if they did not get the correct materials.

27. a. Interphase is the period between cell divisions. It would be impossible for the cell to divide during interphase.

28. e. The first four choices are all said to occur during prophase. Only the last choice is not talked about. It would be very bad if the cell membrane did disappear because all of the cellular material would then scatter.

29. c. The spindle fibers form from the aster. The fibers attach to the chromosomes.

30. c. The last two paragraphs list some of the differences between plant and animal cell division. There is no mention of an animal cell making a cell plate, but it is stated that plant cells make one.

31. b. Telophase is the opposite of prophase.

32. d. A cell must get only one set of chromosomes from each parent to keep the number of chromosomes from doubling in each generation. A cell with only one set of chromosomes is said to be haploid.

33. a. Only the first choice is true. Only one egg cell is produced in meiosis and it is haploid. All four of the sperm cells produced are capable of fertilizing the egg.

34. a. The paragraph states that Gregor Mendel is also called the "father of genetics".

35. a. A dominant trait is one that shows itself even when another trait is also present. In the case shown, the tall gene shows up even though the short gene is present.

36. d. A heterozygous animal is one that has unlike alleles for a particular gene. Animals with like alleles are called homozygous.

37. c. From the diagram it is easy to see that hybrid tan mice will have both white and tan babies. Since the babies are both colors (white and tan) the babies must have more than one phenotype and genotype.

38. e. You must be very careful when you answer this question. At first glance c and d both appear as good choices. Upon closer examination, you can see that the answers say mitosis instead of meiosis. That leaves us with answer e: genes can be inherited independently.

39. a. The paragraph states that because of improvements in microscopes and chemical analysis, there are at least three major classifications of organisms. Going one step further from there, it would be safe to assume that microscopes and chemical analysis are used to identify organisms.

40. a. Viruses are classified by the type of host they infect.

41. d. Amoeba move by pseudopods.

42. d. Algae make their own food by photosynthesis.

43. d. Lichens maintain a symbiotic relationship with algae, which supply the food.

44. b. Conifers stay green year round.

45. a. The leafy part of a fern is called a frond.

46. d. Looking back at the photo of plant leaves, we can see that the monocot leaf has veins that run next to each other while the dicot has netted leaves.

47. b. Plants get water and minerals from the soil.

48. c. Phloem transports materials downward in the plant. One of the main things it transports is food. Glucose is a type of food that is produced in the leaves.

49. e. This is a trick question because both water and minerals are transferred upward in a plant. Therefore, you must choose both and not pick a or b. Remember to always pick the best answer.

50. d. The pith comprises the centermost part of woody stems.

51. c. The regulation of the flow of water and carbon dioxide is controlled by holes in the leaves called stomata.

52. a. The pistil is the female part of the plant. The ovary and the style are both parts of the pistil.

53. b. Aves, or birds, are the only vertebrate listed. The other two choices, mollusks and echinoderms, are both invertebrate.

54. b. Sperm develop best at temperatures lower than body temperature. Therefore they develop outside the body to keep cool.

55. c. Sperms grow tails called flagella for propulsion.

56. b. If an egg is not fertilized menstruation will occur.

57. a. A fertilized egg is called a zygote. It implants itself in the uterus.

58. c. The best answer here is platelets. When you cut yourself the platelets rupture to start the clotting process. They cause a fibrous net to form, which traps the red blood cells. When the red blood cells dry a clot is formed. Even though red blood cells make up part of the clot, if there are no platelets to form the net, they will simply flow freely from the wound.

59. b. While most veins carry deoxygenated blood back to the heart, the pulmonary vein carries blood from the lungs to the heart, which has a high oxygen concentration.

60 a. The other choices are all controlled in part by the pituitary gland, which is found in the brain. The pancreas controls sugar. The pancreas produces insulin, which controls the level of sugar in the

blood. People who cannot produce insulin are called diabetics. People with a low insulin production level can control the disease by changing their diet. However, people with more severe cases must get insulin through shots.

61. a. While glands can secrete chemicals, which have an effect on the nerves, they are not part of the nervous system.

62. c. Even though the failure of any organ is bad, many people live for a long time even when they are in total kidney failure. They must go in for dialysis on a regular basis. The machines take over the functions of the kidneys. Therefore people are able to stay on the waiting list a longer time for a kidney. While they are waiting they are capable of leading semi-normal lives.

63. b. This is the only choice that lists both major types of muscle. Choices a, c, and d all list the one kind of muscle found in a particular type. Choice e lists the two kinds of involuntary muscle.

64. e. Answers a, b, and d are all backwards. Choice c is incorrect because it is the level of carbon dioxide that causes you to breathe, not the level of oxygen.

65. a. The correct flow of food is from the mouth, down the esophagus, to the stomach. From the stomach it flows into the small intestines where most of the nutrients are absorbed. Finally the food enters the large intestine where water is reabsorbed before the rest of the wastes are eliminated.

After you have checked your answers to the questions found throughout the chapter, you are ready to take the chapter test.

Now that you have completed your review for the biology-type questions you will find on the GED Science Test, let's try a few sample GED questions. Remember on the actual GED Science Test, each question or group of questions will have a reading passage or a chart. You should be able to correctly answer the questions based on what you read or are able to infer from the material presented.

CHAPTER TEST - BIOLOGY

Questions 1-3 refer to the following chart:

VITAMIN	NAME	PREVENTS	FOOD FOUND IN
A	CAROTENE	NIGHT BLINDNESS DRY SKIN	YELLOW FRUITS AND VEGETABLES, EGG YOLKS AND LIVER
B-1	THIAMINE	BERI BERI, MENTAL DISORDERS	BRAIN, LIVER, WHOLE GRAINS
B-2	RIBOFLAVIN	SKIN DISORDERS	GRAIN, EGGS, MILK, LIVER
B-6	PYRIDOXINE	SKIN DISORDERS, DERMATITIS	GRAIN, LIVER, FISH, KIDNEY
B-3	NIACIN	PELLAGRA, DIGESTIVE DISORDERS	GRAIN, LIVER, MEAT YEAST
	BIOTIN	MUSCLE PAINS WEAKNESS	EGG YOLK, PRODUCED BY BACTERIA IN THE INTESTINES
	FOLIC ACID	ANEMIA	LIVER AND LEAFY VEGETABLES
	PANTOTHENIC ACID	CARDIOVASCULAR NEURAL DISORDERS	MANY FOODS
B-12	CYANOCOBALAMIN	ANEMIA	LIVER, EGGS, FISH, AND MILK
C	ASCORBIC ACID	SCURVY	CITRUS FRUITS, TOMATOES
D	CALCIFEROL	RICKETS	FISH OIL, MILK, SUN ON THE SKIN
E	TOCOPHEROL	ANEMIA	GREEN VEGETABLES, FISH OIL
K	PHYLLOQUINONE	LACK OF BLOOD CLOTTING	GREEN VEGETABLES

116

Question 1:

Which of the following vitamins if absent would cause you to bleed to death from a small wound?

    a.   vitamin C
    b.   vitamin K
    c.   vitamin B-12
    d.   vitamin D
    e.   vitamin B-2

Question 2:

Which of the following foods is not a source of vitamin B-2?

    a.   fish
    b.   grain
    c.   milk
    d.   liver
    e.   eggs

Question 3:

Which is another name for vitamin D?

    a.   Ascorbic Acid
    b.   Tocopherol
    c.   Pantothenic Acid
    d.   Niacin
    e.   Calciferol

Question 4:

Trees are always making new xylem. It is composed of dead cells that form hollow tubes. It is responsible for bringing the water from the roots to the leaves. The rate at which new xylem is formed depends on the growing conditions. During the winter when growth is slow rings are formed. Which of the following is true?

    a.   The thickest tree is always the oldest.
    b.   The tallest tree is always the oldest.
    c.   Trees always grow toward the south.
    d.   Trees in good growing conditions get thicker faster.
    e.   Xylem brings sugar down to the roots.

Question 5:

The surface of human blood can contain two main antigens, proteins that antibodies attack. They are called A and B. If both antigens are missing the blood is labeled O. Both antigens are controlled by a single gene. Neither one is dominant over the other, so if both alleles (possible traits of the gene) are present the blood would test AB. If a child has a father with an AB blood type, which of the following blood types could the child not have?

    a.   O
    b.   B
    c.   A
    d.   AB
    e.   Cannot be determined from the given information

Question 6:

All cells in an organism have the same genetic information. Cells look and act differently depending on the information they use. Which of the following is true?

    a.   Only cells in the pancreas contain the genetic information to make insulin.
    b.   A heart cell has the information to make insulin.
    c.   Brain cells contain the most genetic information.
    d.   Skin cells contain the most genetic information.
    e.   Brain cells contain the least genetic information.

Question 7:

Organisms in a similar niche (place in the environment) develop similar structures. These structures can develop from parts that are vastly different in an organism's close relatives. For example, the wings of a fly and the wings of a bat have very different evolutionary origins. The structure of a bat wing is analogous in humans, a much closer relative than the fly, to the hand. Which of the following is most likely another example of this type of development?

    a.   bat wings and bird wings
    b.   duck feet and frog feet
    c.   human hair and animal fur
    d.   dolphin fins and fish fins
    e.   airplane wings and bird wings

Question 8:

Wine is made when juice is fermented. Yeast is added to the juice and it breaks down the sugar in the juice to form alcohol. In a dry wine the yeast converts all the sugar to alcohol, but in a sweet wine the alcohol level reaches a level high enough to kill the yeast before it has used up all the sugar. Choose the best answer to explain the yeast population in the following graph for a dry wine.

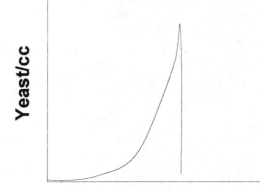

a. The yeast population grew quickly and then died off because the alcohol concentration got too high.
b. The yeast population grew quickly and then leveled off.
c. The yeast population stayed the same the whole time.
d. The yeast population grew and then died off when the sugar ran out.
e. The yeast population decreased as the alcohol concentration grew.

Question 9:

Which of the following is not true about wine?

a. Yeast converts sugar into alcohol.
b. Dry wines contain excess sugar.
c. The sugar content is a factor in determining if a wine is sweet or dry
d. Alcohol in certain concentrations kills yeast
e. Yeast is very important in the wine making process.

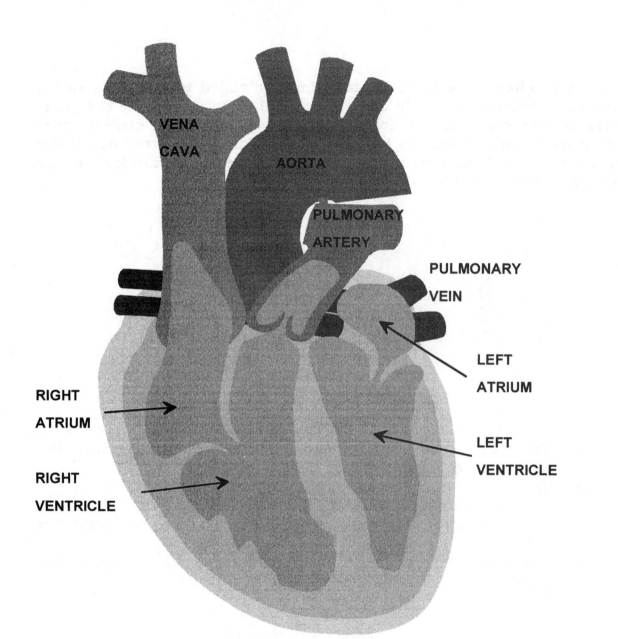

VENA CAVA

AORTA

PULMONARY ARTERY

PULMONARY VEIN

RIGHT ATRIUM

RIGHT VENTRICLE

LEFT ATRIUM

LEFT VENTRICLE

Question 10:

What are the two major arteries leaving the heart?

    a. pulmonary artery and aorta
    b. aorta and pulmonary vein
    c. left ventricle and right atrium
    d. vena cava and pulmonary artery
    e. vena cava and pulmonary vein

Question 11:

What keeps the blood from flowing backwards in the heart?

  a. the contraction of the other chambers
  b. the aorta
  c. valves
  d. doors
  e. blood flows in both directions

Question 12:

Animals with spinal chords belong in the phylum chordata. There are two major subphylum within the chordates, those with backbones and those without backbones. The vertebrates, animals with backbones, are composed of some of the most recognizable classes of animals. They include fish, amphibians, reptiles, aves (birds) and mammals. Which of the following is not true?

  a. the mallard duck is a member of the vertebrates
  b. fish are a type of invertebrate
  c. all mammals are also vertebrates
  d. all vertebrates are chordates
  e. fish are a member of the chordates

Question 13:

XYZ company is doing an experiment to find out which of their fertilizers works best on pea plants. Which of the following would be the best control group for the experiment?

  a. A group of pea plants that gets no sunshine.
  b. A group of pea plants, which gets fertilizer but no water.
  c. A group of pea plants, which gets double the amount of fertilizer.
  d. A group of bean plants, which gets the same amount of water and fertilizer as the rest.
  e. A group of pea plants that gets the same light and water as the rest, which gets no fertilizer at all.

	Week 1	Week 2	Week 3	Week 4
No fertilizer	4"	5"	6"	8"
Fertilizer 1	4"	6"	9"	12"
Fertilizer 2	4"	5"	6"	8"
Fertilizer 3	4"	6"	8"	10"

Question 14:

Which fertilizer is a waste of money?

        a.   fertilizer 1
        b.   fertilizer 2
        c.   fertilizer 3
        d.   all of them work well
        e.   none of them work at all

Question 15:

If all the fertilizers cost the same amount, which fertilizer is the best value for the dollar?

        a. fertilizer 1
        b. fertilizer 2
        c. fertilizer 3
        d. all of them would have the same value
        e. cannot be determined from the given information

The following data represents data obtained in the determination of dominance and recessive in a flower. Each tally mark represents one offspring of that color.

	RR	Rr	rr
Tally marks	~~IIII~~ ~~IIII~~ ~~IIII~~ ~~IIII~~ IIII	~~IIII~~ ~~IIII~~ ~~IIII~~ ~~IIII~~ ~~IIII~~ ~~IIII~~ ~~IIII~~ ~~IIII~~ ~~IIII~~ ~~IIII~~	~~IIII~~ ~~IIII~~ ~~IIII~~ ~~IIII~~ ~~IIII~~ 1
Totals	24	50	26

Genotype = RR : Rr : rr   Phenotype= Red : White
           24 : 50 :26   74 : 26

Question 16:

What percentage of 100 offspring may be predicted to show the dominant trait?

    a.   0%
    b.   24%
    c.   74%
    d.   50%
    e.   26%

Question 17:

What are the chances of getting a white trait from the cross of Rr and Rr?

    a.   3 to 1
    b.   1 to 4
    c.   2 to 5
    d.   1 to 3
    e.   1 to 1

Question 18:

If additional data were collected, what may be predicted?

    a.   The genotype ratio would be the same.
    b.   The majority selection would be "rr".
    c.   The majority selection would be "RR".
    d.   The results would be totally unpredictable.
    e.   More data would be needed.

Question 19:

A common form of reproduction in animals is sexual reproduction. The parents of the offspring are both diploid (have two copies of each chromosome) and they produce some special cells that are haploid. One haploid cell from the mother combines with one haploid cell from the father. This forms a new diploid offspring. Which of the following is not true?

    a.   Diploid animals only have one copy of each gene.
    b.   Offspring get information from both parents.
    c.   Offspring have the same number of chromosomes as the parents.
    d.   Both parents have the same number of chromosomes.
    e.   Parents must make special cells to reproduce.

Question 20:

In Darwin's theory of evolution, he stated that organisms evolve based on their fitness to produce viable (young that are able to reproduce) offspring. Animals that produce more viable offspring than other members of their species will cause their characteristics to show up in future generations. Those that do not produce viable offspring do not pass along their characteristics. Which of the following is not an example of "survival of the fittest?"

a. Antelope A is faster than antelope B. Antelope B gets caught and eaten by a lion before it has offspring while antelope A dies of old age after having many offspring.

b. Reindeer A has a thicker coat than reindeer B. During the winter (when they are young) it is especially cold and reindeer B dies before mating. Reindeer A grows to be the leader of the herd and has many offspring.

c. Bat B has a slight refinement in its echolocation. It is able to catch many more bugs than the average bat. Its children also are able to catch more bugs and have a 75% higher survival rate.

d. In an area, red tail hawks and peregrine falcons compete for the same food source. The falcons being smaller and quicker are able to catch more food and finally the hawks die off.

e. A type of moth is normally light gray to blend in on tree bark, however because humans in the area are burning a lot of coal all the trees are covered with soot. To begin with, only a small portion of the moth population is black, however they blend in much better on the soot covered trees. After a few years, black becomes the most common color for moths in the area.

Question 21:

Photosynthesis is the process by which plants store energy. Plants take the energy contained in sunlight and by using chlorophyll and other enzymes make sugar. Sugar is the basic building block of all food. Multiple sugars can be joined together to make starch. If sugar is joined with glycerol it makes a fatty acid. Finally if sugar is joined to an amine group an amino acid is formed. Which of the following can plants do that animals cannot?

a. convert starch to sugar
b. convert sugar to amino acids
c. produce sugar
d. convert sugar to fat
e. convert fat to sugar

Question 22:

Where do plants get the energy to make sugar?

    a.   from starch
    b.   from sunlight
    c.   from chlorophyll
    d.   from enzymes
    e.   from storage

Question 23:

Many scientists believe that when Earth first formed it was lifeless. The first life to form was found in the oceans. Slowly life became more complicated and multi-cellular organisms formed. The next great age was the age of invertebrates, followed by the age of fish. The next age is the first animals to live at least part-time on land, amphibians. Following the amphibians were the reptiles. Finally, along came the mammals. Which of the following places the great ages from youngest (most recent) to oldest?

    a.   amphibians, fish, reptiles
    b.   amphibians, reptiles, mammals
    c.   fish, amphibians, reptiles
    d.   amphibians, no life, reptiles
    e.   mammals, fish, no life

ANSWERS - CHAPTER TEST BIOLOGY

1. b. In this type of question you must simply look down the columns listing the things the vitamins prevent and chose the vitamin important in blood clotting. Vitamin K is important in blood clotting. Without it, even small cuts can be very dangerous.

2. a. Here you are required to find vitamin B-2 and then follow the row over to find in which foods it is found. The only choice not listed is fish.

3. e. Here you must first find vitamin D and then follow the row over to the vitamin names. Calciferol is the name of vitamin D.

4. d. This question is a little tougher than simple chart reading; you must do some analysis. The first answer looks correct; however, the paragraph states trees form new xylem at different rates, based on growing conditions. Therefore, a younger tree in good growing conditions might soon grow bigger than an older one in poorer growing conditions. Likewise, the tallest tree is not always the oldest. Answer e, states the opposite of what is in the paragraph. The paragraph states xylem takes the water from the roots to the leaves. Nothing in the paragraph should make you pick answer c. When you take the test always make sure you eliminate answers you know are wrong before you make your best guess at the correct answer. Finally, answer d is supported by the paragraph. It states new xylem is formed faster when the tree is in good growing conditions, which means the trees will get thicker, faster.

5. a. Again, you would need to do a little analysis for this problem. If the father is AB, he would need to pass on either an A or B gene to his children. If a person has at least one A or B gene, he cannot have an O blood type.

6. b. The paragraph states all cells in an organism contain the same information. If this is true, heart cells must contain the information to produce insulin, but do not use it.

7. d. Here you are required to first understand the presented concept and then apply it to come up with another example. First you have to look at the animals and find ones which are not very close evolutionarily. Human hair and animal fur are too similar for what we are trying to find out. Likewise, bat wings and bird wings derive from the same evolutionary origins. The same goes for duck feet and frog feet. Choice e is definitely wrong because it compares an animal with an inanimate object. This leaves us with our correct answer: dolphin fins and fish fins. If you were to cut open a dolphin fin, you would find a bond structure similar to the human hand.

8. d. There are two reasons why the yeast population could die off. The first is because the alcohol concentration went too high. The second is the yeast used up all the sugar. In the case of a dry wine, all the sugar is used up.

9. b. This is the opposite of what the paragraph states. Wines with excess sugar are sweet.

10. a. If you are unsure of the correct answer, look at the valves in the heart. If blood were to try and flow from the heart into the pulmonary vein or into the vena cava, the valves would close and cut the blood flow off.

11. c. Valves keep the blood from flowing backward in the heart.

12. b. Fish are listed as members of the vertebrates.

13. e. A good control group gets everything the other groups get, except for the product you are testing. Therefore, choice b is the best.

14. b. A fertilizer that is a waste of money would show the same or worse results as the control group. Fertilizer 2 has the same results as the control group, so the best answer is b.

15. a. If all the fertilizers cost the same amount of money, the fertilizer getting the best results would return the most value. The best answer is Fertilizer 1, or answer a.

16. c. An offspring that shows the dominant trait would have to have at least one copy of the dominant gene. The results show there are twenty-four with both dominant genes and fifty others with just one dominant gene. Adding the two together, the total is seventy-four. There were a total of one hundred offspring, so 74% of the offspring received at least one dominant trait.

17. d. This is very easy to see if you use a Punnett square.

	R	r
R	RR	Rr
r	rR	rr

The only flowers which will have white flowers are the ones with both little r's from the cross. Therefore, you would expect to get only one white flower for every three red flowers.

18. a. The genotype of the cross would be expected to remain the same, no matter how many offspring are produced. The larger the sample, the closer to the expected results you would get.

19 a. The paragraph states diploid animals have two copies of each gene. The other answers are all true.

20. d. Example d compares two different species. Survival of the fittest compares animals within the same species and their ability to pass on their genes to their offspring.

21. c. Animals cannot produce their own food, they must eat it. Animals can convert sugar to fats and amino acids, as well as convert amino acids and fats back to sugar.

22. b. Plants get the energy to make sugar from sunlight.

23. e. Answer b is the only choice listing the ages from youngest to oldest. Answer e lists the ages from oldest to youngest. Always make sure you answer the question being asked.

SECTION 1: STRUCTURE AND COMPOSITION OF THE ATOM

Chemistry involves the study of the basic particles that make up everything we see and how these particles are joined together. The basic building block of all matter is the atom. Matter is everything around us. Even we, ourselves, are made up of matter. Matter includes all solids, liquids, and gases.

Question 1:

What is the basic building block of matter?

     a. atoms
     b. gases
     c. liquids
     d. solids
     e. particles

ATOMS

Atoms are made of three large particles. These particles occupy specific locations in the atom.

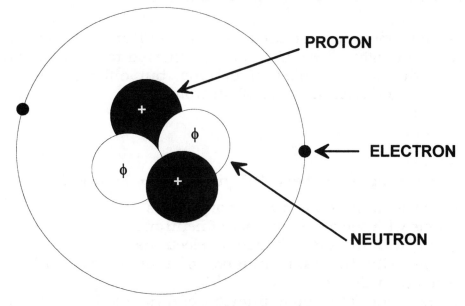

ATOMIC WEIGHT

Protons and **neutrons** are located in the **nucleus** (the center) of the atom. Each of these particles weighs one atomic mass unit (1 amu). Together, they make up the **atomic weight** (how much the atom weighs) of the atom.

Atomic weight may accompany the chemical symbol in certain problems. It is the number written as a superscript to the right or upper left of the chemical symbol.

EXAMPLE: Pt^{195} or ^{195}Pt

Ga^{70} or ^{70}Ga

Question 2:

What makes up the nucleus?

   a. protons
   b. neutrons
   c. electrons
   d. a and b
   e. all of the above

ISOTOPES

Atoms having the same number of protons are said to be the same element. For example, hydrogen always has one proton. Although the <u>atoms of the same element contain the same number of protons</u>, the number of neutrons may vary.

Therefore, uranium may be expressed as ^{238}U, ^{235}U or ^{234}U, depending upon the specific atomic weight. These U atoms are referred to as **isotopes**. Isotopes are atoms with the same number of protons, but different atomic weights because they have a different number of neutrons.

Question 3:

Which of the following is the best definition for an isotope?

   a. atoms with the same number of protons
   b. atoms with the same number of neutrons
   c. atoms with the same number of electrons
   d. atoms with the same number of protons but with a different number of electrons
   e. atoms with the same number of protons but with a different number of neutrons

Many atoms have isotopic forms.

ISOTOPES OF HYDROGEN

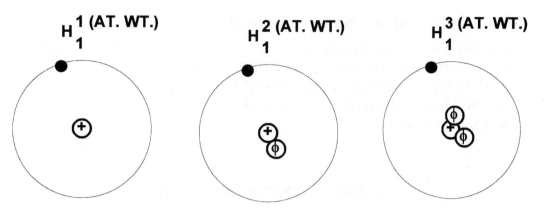

^1_1H (AT. WT.)

^2_1H (AT. WT.)

^3_1H (AT. WT.)

All of these are hydrogen because they have one proton and electron. But, they all are isotopes of hydrogen because they are different atomic weights of the same element.

ATOMIC WEIGHTS OF CARBON

$^{12}_6C$

6 Protons
6 Neutrons

$^{12}_6C$

6 Protons
8 Neutrons

$4e^-$

$2e^-$

$4e^-$

$2e^-$

All of the above are carbon because they have 6 p^+ and 6 e^-. But, they all are isotopes of carbon because they have different atomic weights of the same element.

Question 4:

Which of the following is true of isotopes?

    a. they all react differently
    b. they have a different number of protons
    c. they have a different number of electrons
    d. they have a different number of neutrons
    e. they explode

ATOMIC NUMBER PROTONS

The number of protons in the nucleus determines the atomic number of the atom. Each element (group of similar atoms) has its own atomic number. When this number accompanies the chemical symbol, it is placed to the lower left of the symbol.

Atomic Weight 1 $_1$H   Atomic Weight 7 $_3$Li   Atomic Weight 238 $_{92}$U
Atomic Number 1   Atomic Number 3   Atomic Number 92

Question 5:

What is the best definition for atomic number?

    a. the number of electrons in the atom
    b. the number of protons in the atom
    c. the number of neutrons in the atom
    d. the number of electrons and protons in the atom
    e. the number of protons and neutrons in the atom

ELECTRICAL CHARGES OF THE ATOM

The number of protons in the nucleus determines the number of electrons present in an **atom**. Atoms are electrically **neutral**. For every proton there is an electron. The positive electrical charge of the proton is cancelled by the negative electrical charge of the electron. (In an atom, (+) charges = (-) charges, leaving the atom in a neutral state electrically).

Question 6:

How many electrons would oxygen have if it has eight protons?

    a. four
    b. eight
    c. sixteen
    d. twenty-four
    e. not enough information given

ELECTRONS

Electrons are located outside the nucleus at sites called rings/orbits/shells/levels (take your choice, they all mean the same site).

LEVELS

An atom may contain up to seven levels.

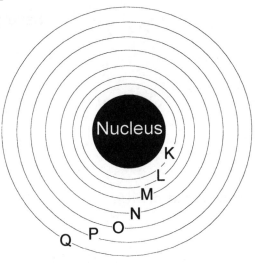

Naturally, these levels exist only if there are electrons present at that level.

EXAMPLES:

$^{1}_{1}H$

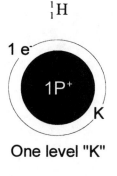

One level "K"

$^{35}_{17}Cl$

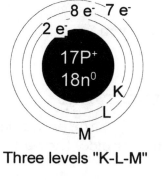

Three levels "K-L-M"

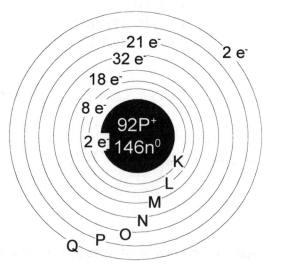

Seven levels "K-L-M-N-O-P-Q"

Question 7:

Electrons are located in

    a. rings.
    b. shells.
    c. levels.
    d. orbits.
    e. rings, shells, levels, and orbits.

SECTION 2: NUMBER OF ELEMENTS

There are ninety-two elements naturally occurring in nature. There are also a number of man-made atoms, but these disintegrate quickly back to the natural ninety-two.

CHEMICAL SYMBOLS

Each element has its own chemical symbol. This symbol may be one capital letter or a capital letter followed by a lower case letter.

EXAMPLES: H   hydrogen   C   carbon

He   helium   Ca   calcium

Here is a list of common chemical symbols you might be expected to know for the GED examination:

CHEMICAL SYMBOL	ELEMENT	CHEMICAL SYMBOL	ELEMENT
1. Al	Aluminum	11. Na	Sodium
2. Ca	Calcium	12. S	Sulfur
3. C	Carbon	13. U	Uranium
4. Cu	Copper	14. Zn	Zinc
5. Fe	Iron	15. Hg	Mercury
6. Pb	Lead	16. Br_2	Bromine
7. Mg	Magnesium	17. N_2	Nitrogen
8. P	Phosphorus	18. O_2	Oxygen
9. K	Potassium	19. H_2	Hydrogen
10. Si	Silicon	20. Cl_2	Chlorine

Elements 1-14 exist naturally as solids; 15 and 16 exist as liquids. Numbers 17-20 exist as diatomic gases.

Question 8:

Which of the following are solids at room temperature?

I   OXYGEN

II   POTASSIUM

III   CARBON

a. I only
b. II only
c. I and III only
d. II and III only
e. all of the above

SECTION 3: DIATOMIC MOLECULES AND COMPOUNDS

The diatomic gases have a **subscript** written to the lower right of the chemical symbol (O_2, H_2, N_2, and Cl_2). These gases exist in their free state as a pair of atoms and referred to as a **diatomic molecule** (two atoms forming one molecule).

$$O_2 \qquad Cl_2$$

COMPOUNDS

Given the right circumstances, elements, compounds, or both may react with each other to form new products. The products formed may be compounds or released elements and will have their own special chemical and physical properties.

Question 9:

When two like atoms form a molecule, what are they called?

        a. a joint
        b. a joining
        c. bicompound molecules
        d. diatomic molecules
        e. unstable compounds

SECTION 4: CHEMICAL REACTIONS

When two **elements** combine, the properties of the new **product** may be different. Look at the example of what happens when hydrogen and oxygen combine to form the new **compound**, water.

TWO ELEMENTS	hydrogen + oxygen $\Rightarrow$	water
SYMBOLS	$H_2 +$ $O_2 \Rightarrow$ (unbalanced)	H_2O
PHYSICAL PROPERTIES	gases, odorless, tasteless, colorless	liquid, odorless, tasteless, colorless
CHEMICAL PROPERTIES	very supports explosive combustion	NOT explosive and does NOT support combustion

Question 10:

Which of the following is true?

   a. The products of a reaction always have the same properties as the reactants.
   b. The products of a reaction always have a blend of the properties of the reactants.
   c. There are always more products than reactants.
   d. The products will have properties independent from those of the reactants.
   e. The reactants always explode.

Now, look at what happens when two compounds are combined. This reaction is called a **double replacement reaction**:

TWO COMPOUNDS

sodium hydroxide + hydrochloric acid $\Rightarrow$ sodium chloride + water

When this is written as an equation using the symbols that represent the compounds it would look like this:

SYMBOLS $Na(OH) + HCl \Rightarrow NaCl + H_2O$

IMPORTANT POINTS FOR YOU TO KNOW WHEN WORKING WITH ELEMENTS, COMPOUNDS, AND REACTIONS:

- (OH) with or without brackets, usually denotes a **base**.

- H beginning a chemical formula usually denotes an **acid**.

- an acid + a base combine to form a salt and water

Using this information with the previous equation, we see that Na (OH) is a base, HCl is an acid and that they combined to form NaCl, which is a salt, and H_2O, which is water.

$$BASE \ + \ ACID \Rightarrow SALT \ + \ WATER$$

$$Na\,(OH) + \ HCl \Rightarrow NaCl \ + \ H_2O$$

When a compound and an element are combined to form a new product the process is called a **single replacement reaction**.

Look at the following example of a single replacement reaction.

$$Sodium\ Chloride + Fluorine \Rightarrow Sodium\ Fluoride + Chlorine$$

Using the symbols that represent these chemicals, the equation would look like this:

SYMBOLS $\ \ NaCl \ + \ F_2 \Rightarrow NaF \ + \ Cl_2$

(unbalanced)

Question 11:

Which of the following is a double replacement reaction?

I. $LiOH + HCl \Rightarrow LiCl + H_2O$

II. $H_2 + O_2 \Rightarrow 2H_2O$

III. $C + O_2 \Rightarrow CO_2$

a. I only
b. II only
c. III only
d. II and III only
e. all of the above

THE LAW OF CONSERVATION OF MATTER

At this point, it would be wise to study the **Law of Conservation of Matter**. Simply stated: <u>matter can neither be created nor destroyed</u>.

Look at the following reaction formula, $H_2 + O_2 \Rightarrow H_2O$.

It is true that hydrogen and oxygen combine to form water. However, the <u>correct amount</u> of H_2 must combine with the <u>correct amount</u> of O_2 in order to produce H_2O.

The "CORRECT" amount of reactants in a reaction (the ingredients are called reactants) is the <u>least</u> amount that allows the reaction to proceed. H_2O is the proper formula for water. H_2 and O_2 is the proper way to represent the reactants. Unfortunately, in the overall formula there is <u>NO</u> equality.

H_2 two atoms of hydrogen (one molecule)

O_2 two atoms of oxygen (one molecule)

H_2O two atoms hydrogen, one atom oxygen (one molecule)

$$H_2 + O_2 \Rightarrow H_2O$$
$$2 : 2 \qquad 2:1$$

In the reaction as written, one oxygen has been "lost". The Law of Conservation of Matter states this cannot happen. In order to remedy the situation and comply with the Law, a balance must be struck on each side of the arrow.

Suppose we use two molecules of H_2 (four atoms) on the left side of the arrow and two molecules of water on the right side of the arrow.

$$\underline{2}H_2 \qquad + O_2 \qquad \Rightarrow \quad \underline{2} \ H_2O$$

$$4 \text{ H ATOMS} + 2 \text{ O ATOMS} \Rightarrow 4 \text{ H ATOMS} + 2 \text{ O ATOMS}$$

This balances our formula - the amount of matter on the right side of the arrow equals the amount of matter on the left side of the arrow:

	H_2	$O_2 \Rightarrow H_2O$	
UNBALANCED	1 molecule 2 atoms	1 molecule 2 atoms	1 molecule water = 2 atoms hydrogen and 1 atom of oxygen
BALANCED	$2H_2$ 2 molecules of <u>2 atoms each</u>	$+ O_2 \Rightarrow$ 1 molecule <u>of 2 atoms</u>	$2 H_2O$ 2 molecules <u>of water</u>
	4 atoms (2x2) of hydrogen	2 atoms (1x2) of oxygen	2x2 hydrogen 4 atoms hydrogen
			2x1 oxygen <u>2 atoms of oxygen</u>

Question 12:

Why is it necessary to balance equations?

       a. so that you are not creating matter
       b. to make everything easier
       c. to make everything harder
       d. to produce more oxygen
       e. to eliminate oxygen

SECTION 5: COVALENT BONDING

When atoms form bonds they can either give up, take away, or share electrons. The sharing of electrons is called **covalent bonding**.

EXAMPLE: $2 H_2 + O_2 \Rightarrow 2 H_2O$

    $4 H + 2 O$ will produce 2 molecules of H_2O

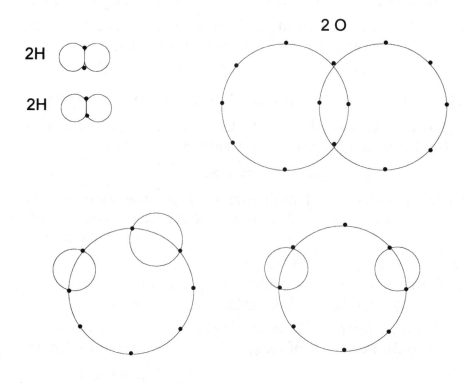

The two diatomic molecules of hydrogen combine with the one diatomic molecule of oxygen. During the reaction, the diatomic molecules of hydrogen and oxygen are broken apart and reassembled.

Question 13:

When hydrogen and oxygen form a molecule, which of the following is true?

        a. They share the electrons.
        b. The oxygen takes electrons from the hydrogen.
        c. The hydrogen takes electrons from the oxygen.
        d. Nothing happens to the electrons.
        e. The oxygen takes protons from the hydrogen.

SECTION 6: IONIC BONDING

Another kind of chemical bonding is the **ionic bond**. In this type of bond the electrons are moved from one atom to another. Therefore, an ionic bond is one in which an atom gains or loses an electron.

EXAMPLE:

$$2\,Na \quad + \quad Cl_2 \rightarrow \ 2\,NaCl$$

(to simplify, we'll work with only one atom of Na, Cl, and one molecule of NaCl)

EXAMPLE:

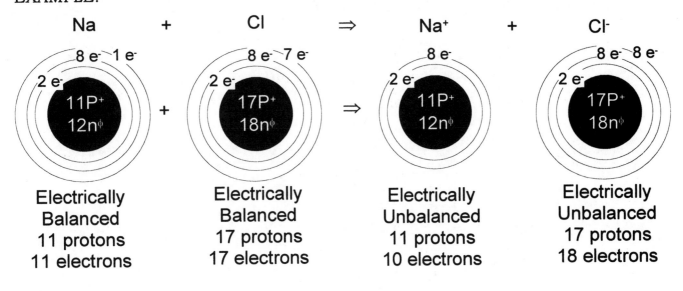

| Na | + | Cl | $\Rightarrow$ | Na$^+$ | + | Cl$^-$ |

Electrically
Balanced
11 protons
11 electrons

Electrically
Balanced
17 protons
17 electrons

Electrically
Unbalanced
11 protons
10 electrons

Electrically
Unbalanced
17 protons
18 electrons

Na$^+$ and Cl$^-$ can be written with + and - indicating they are charged (electrically UNbalanced). They are now called IONS because they no longer fit the definition of an atom.

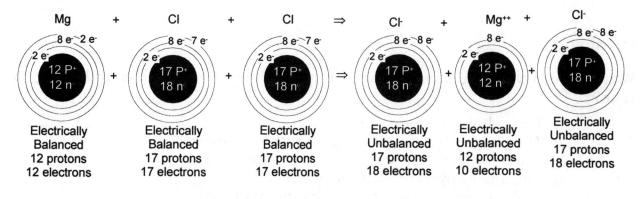

| Mg | + | Cl | + | Cl | ⇒ | Cl$^-$ | + | Mg^{++} | + | Cl$^-$ |

Electrically Balanced
12 protons
12 electrons

Electrically Balanced
17 protons
17 electrons

Electrically Balanced
17 protons
17 electrons

Electrically Unbalanced
17 protons
18 electrons

Electrically Unbalanced
12 protons
10 electrons

Electrically Unbalanced
17 protons
18 electrons

Question 14:

What is the product of an ionic reaction?

        a. a molecule
        b. a compound
        c. some ions
        d. many molecules
        e. cannot be determined

You have now completed your review of chemistry. Before you take the chapter test, check your answers to the chemistry questions that appeared in the chapter.

ANSWERS - CHAPTER QUESTIONS

1. a. The paragraph states the basic building block is the atom and that it makes up all matter.

2. d. The paragraphs state both protons and neutrons are found in the nucleus, and that together they make up the atomic weight of an atom.

3. e. The last paragraph gives the best definition of an isotope. It says isotopes have the same number of protons, but different atomic weights because they have different numbers of neutrons.

4. d. The examples reinforce what we saw in the previous paragraphs, that isotopes have a different number of neutrons.

5. b. The passage states the atomic number is the number of protons in the nucleus.

6. b. The paragraph states atoms have as many electrons as protons. If oxygen has eight protons, it would also have eight electrons.

7. e. The first paragraph states electrons are found in rings, levels, shells, and orbits, and that they all mean the same thing.

8. d. Looking at the chart provided we can see that oxygen is a diatomic gas, potassium is a solid, and carbon is a solid. That makes d your best answer.

9. d. The first paragraph tells about diatomic molecules. It says that they are two like atoms that are joined together. When unlike atoms are joined, the product is called a compound.

10. d. Looking at the reaction in the given example, the only answer that fits is the products will have properties independent from the reactants. In the example you go from having two gases to having a liquid. That is definitely not a blending of properties. We ended up with fewer products than reactants, and it is not stated that all reactants must explode.

11. a. A double replacement reaction is when two reactants form two products. The only reaction that does that is the first reaction.

12. a. The paragraphs state the reason it is necessary to balance equations is that you cannot create matter.

13. a. The paragraphs state hydrogen and oxygen form a covalent compound. This means that they share electrons.

14. c. In an ionic reaction, the reactants would combine to make ions.

Review the areas which you answered incorrectly and then take the chapter test.

As you take this short quiz, remember the information you need to answer the question is presented in the paragraph, chart, or diagram. You must simply analyze the information properly.

CHAPTER TEST - CHEMISTRY

Question 1:

It is important for chemists to keep track of all the reactants in a reaction. All the elements in the reactants must end up in the products. This is done by balancing each equation. A chemist looks at each element, for example carbon, and if there are four carbons in the reactants there must be four carbons in the products. This is done for each element until the equation balances.

What must Y equal in order for the following equation to balance?

$$2H_2 + O_2 \rightarrow Y\,H_2O$$

    a.   1
    b.   2
    c.   3
    d.   4
    e.   5

Question 2:

Atoms are named based on the number of protons they contain. The number of protons in an atom is called the atomic number. Protons are found in the nucleus of the atom. Also found in the nucleus are neutrons. Atoms with the same number of protons but different numbers of neutrons are called isotopes. They are called by the same name, but one would weigh more. The weight of an atom is called the atomic weight. Atoms with different atomic weights chemically react the same.

Name	Chemical Symbol	Atomic Number	Most Common Atomic Weight
Hydrogen	H	1	1
Helium	He	2	4
Lithium	Li	3	7
Beryllium	Be	4	9
Boron	B	5	11

Carbon	C	6	12
Nitrogen	N	7	14
Oxygen	O	8	16
Fluorine	Fl	9	19
Neon	Ne	10	20
Sodium	Na	11	23
Magnesium	Mg	12	24

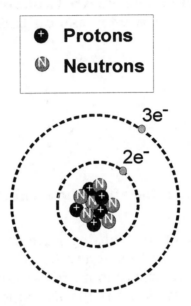

What is the name of this atom?

    a.   Helium
    b.   Lithium
    c.   Boron
    d.   Carbon
    e.   Sodium

Question 3:

How many elements in the table have more neutrons than protons?

    a.   1
    b.   2
    c.   5
    d.   6
    e.   11

Question 4:

Every radioactive element has a half-life. The half-life of an element is the amount of time it takes for half of the material to decay into other elements. For example, the half-life of U^{238} is 4.5 billion years. Other elements have much shorter half-lives; some are measured in parts of a second.

If an element has been around for two half-lives, then only one quarter of the original material is left. If it has been around for three half-lives, then only one-eighth of the original material is left.

Many scientists believe that the Earth is about 4,500,000,000 years old. Approximately how much more U^{238} was on the Earth when it was formed than there is today?

a. One quarter
b. One half
c. The same
d. Twice as much
e. Four times as much

Question 5:

Why must humans be more careful when they dispose of nuclear waste than with other trash?

a. Because some of the half-lives are so long, it takes a long time for the waste to decay to safe levels.
b. Because nuclear waste is more valuable. It must be protected.
c. Because nuclear waste is highly explosive.
d. Because everyone wants nuclear waste.
e. Because most nuclear waste comes from hospitals.

Question 6:

Ions are formed when an atom gains or looses an electron from its neutral state. The most common way this is represented is with a superscript + or - next to the chemical symbol. For example, a sodium ion that has lost an electron would be represented as Na^+, because it would now have a +1 charge. Which of the following would represent an oxygen atom that has gained two electrons?

a. O_2
b. O^+
c. O^2
d. O^{+2}
e. O^{-2}

Question 7:

A diatomic molecule is one where two atoms of the same element form one molecule. Some of the elements that form diatomic molecules are oxygen, hydrogen and chlorine. Which of the following would be a chemical symbol for a diatomic molecule?

a. $2\,Cl$
b. Cl^2
c. O_2
d. HCl
e. OH

Question 8:

Oxygen is able to form the semi-stable compound ozone. When it forms near the earth's surface, for example from car exhaust, it is a pollutant. However, when it forms high in the atmosphere, for example from lightning, it is beneficial. This type of ozone forms a protective layer around the earth. It blocks some of the high energy ultra-violet radiation before it reaches the earth's surface. Unfortunately, some of the pollutants released by humans break ozone back into regular oxygen. If ozone has three atoms of oxygen and oxygen normally occurs as a diatomic molecule, which of the following reactions would represent a balanced equation for the breakdown of ozone to oxygen?

a. $O_2 \rightarrow 3O$
b. $3O_2 \rightarrow 2O_3$
c. $3O \rightarrow 2O_3$
d. $2O_3 \rightarrow 3O_2$
e. $O_3 \rightarrow O_2$

Question 9:

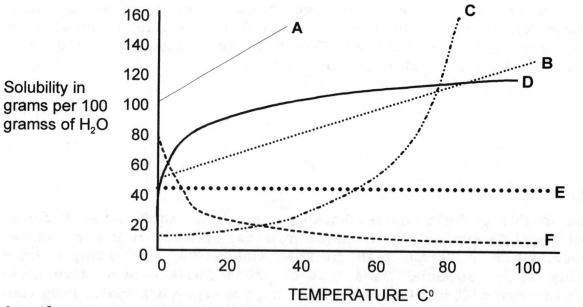

As the temperature increases, the solubility decreases for which substance?

a. A
b. B
c. C
d. D
e. F

ANSWERS - CHAPTER TEST CHEMISTRY

1. b. The equation balances when Y equals 2.

   $$2H_2 + O_2 \rightarrow 2H_2O$$

2. c. Looking at the diagram, you can see there are five protons in the nucleus. From the table, you can see the atom with five protons is boron.

3. d. An atom with more neutrons than protons would have an atomic weight more than double its atomic number. There are six atoms listed with atomic weights more than double their atomic numbers.

4. d. The Earth has been around 4.5 billion years and U^{238} has a half life of 4.5 billion years. There is now one-half the original amount of U^{238}, or originally there was twice as much.

5. a. If it takes 4.5 billion years for U^{238} to degrade by just one-half, you have to be very careful with wastes which contain it because it will remain on Earth for a very long time.

6. e. An atom that gains two electrons would have a –2 charge. An oxygen atom with a –2 charge would be written as O^{-2}.

7. c. A diatomic molecule must have two of the same atom, so immediately you could eliminate choices d and e. Answer a is the representation of two individual chlorine atoms. Answer b is how you would represent a chlorine atom with a positive two charge.

8. d. The reaction must have O_3 breaking down into O_2. Only choices d and e are possible correct answers. Choice e is not balanced. So, d for is left as the correct answer.

9. e. Higher temperature is represented on the right hand side of the bottom line. The only line that is lower on the right than it is on the left is line F.

This completes the review for the GED chemistry questions. Restudy the areas where you are still having difficulty and then go on to your review for the physics questions.

CHAPTER 4 – PHYSICS

We are now going to cover some of the basic principles of physics. Because the test is set up with reading passages about which you will be asked questions, we will have questions as you read through the material.

SECTION 1: MOTION

Sir Isaac Newton formulated the laws of motion. Motion is the result of a force acting on a mass.

First Law of Motion: The two parts of the First Law of Motion are:

1) A body at rest tends to stay at rest

2) A body in motion tends to stay in motion unless acted upon by an outside force.

EXAMPLES:

1) A book on a table will stay there forever unless someone or something applies a force to, causing it to move.

No friction or gravity

Actual

2) A cannon shot would travel in a straight line forever if it wasn't for the friction of the air slowing it down and the force of gravity pulling it to the surface of the Earth.

Question 1:

What is the flight path of a bullet?

    a. A straight line.
    b. A line that curves with the earth.
    c. A line that points out into outer space.
    d. A line that is curved due to gravity.
    e. None of the above.

Question 2:

What happens to a marble that is set on a level desk?

    a. It rolls off.
    b. It rolls in a circle.
    c. It stays where it is placed.
    d. It would follow a curved path off the table.
    e. None of the above.

<u>Second Law of Motion</u>:   Force = mass x acceleration, or F = ma

EXAMPLE:

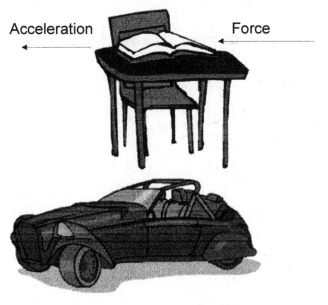

Acceleration   Force

The book on the table in the example has a mass of 1 kg. Some force is applied causing the book to move from its original position. The change from an initial velocity to a new velocity represents acceleration (Ex.: changing speeds in a car from 40 mph to 55 mph. That change is referred to as acceleration). Let us say the acceleration was 2 m/sec. By applying the numbers to the formula we get

$$F = 1 \text{ kg x } 2 \text{ m/sec}$$

$$F = \frac{2 \text{ kg-m}}{\text{sec}} \text{ or 2 nt (Newtons)}$$

The book only moves because a force causes it to accelerate. Obviously, this force would be very negligible if one were to attempt to use it to stop a speeding car.

Question 3:

What would be the acceleration of a 3 kg object if a 9 (kg x m/sec) force is applied to it?

        a. 1/3 (m/sec)
        b. 3 (m/sec)
        c. 9 (m/sec)
        d. 27 (m/sec)
        e. not enough information given

Third Law of Motion: For every action there is an equal and opposite reaction.

EXAMPLE:

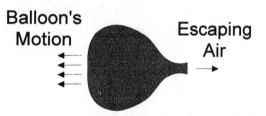

Balloon's Motion   Escaping Air

A balloon is a good example of action-reaction. When releasing a balloon filled with air, the air escapes out the blow hole while the balloon is propelled forward.

Question 4:

If you throw a baseball while you are standing in a row boat, what happens?

        a. The ball moves, but the boat does not.
        b. The boat moves, but the ball does not.
        c. The boat and the ball move the same amount.
        d. The boat and ball move the same direction.
        e. None of these.

Newton is also responsible for our understanding of gravity. The formula for gravity is derived from F = ma. It is:

$$\text{Force of gravity} = \frac{(\text{Gravity constant})(\text{mass})(\text{mass})}{(\text{Distance})^2} \quad \text{to simplify,} \quad F = \frac{G\,M_1\,M_2}{D^2}$$

Examples: 1) Two large masses near each other:

Mass = 2

Mass = 4

distance = 2

M_1 $\longleftrightarrow$ M_2

$$F = \frac{M_1 M_2}{d^2} = \frac{(2)(4)}{2^2} = 2 \text{ gravity units}$$

2) Two small masses near each other:

Mass = 2

Mass = 1

distance = 2

M_1 $\longleftrightarrow$ M_2

$$F = \frac{M_1 M_2}{d^2} = \frac{(1)(2)}{2^2} = \frac{1}{2} \text{ gravity units}$$

To summarize: the greater the distance between the masses, the less gravitational attraction.

By the way, the acceleration due to the force of gravity on earth is 9.8 m/sec² or 32 ft./sec².

Question 5:

The moon is closer to the earth than the sun is to the earth. The moon has a greater impact on the tides through its gravitational pull than the sun has. Which of the following is true?

   a. The moon has a stronger gravitational attraction to the earth than the sun.
   b. The moon is bigger than the earth.
   c. The moon is bigger than the sun.
   d. The moon and the sun are the same size.
   e. There is not enough information given.

153

SECTION 2: THE LEVER

A lever is a bar that is balanced at a point called the fulcrum. There are three classes of levers - First, Second, and Third.

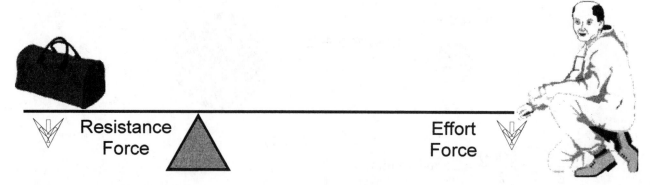

Remember the seesaw when you were young? The lever seen above is that kind of machine. You know that the heavier person always had to sit closer to the middle of the board and the lighter person could balance the heavier person. That's what this lever does.

EFFORT FORCE   **RESISTANT FORCE**

Other examples of first class levers are the juice can opener, crow bar and scissors.

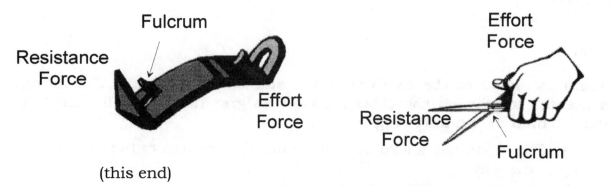

Question 6:

Where does the lighter person have to sit to balance a heavier person on a seesaw?

    a.   At the same distance from the fulcrum as the heavier person.
    b.   Farther from the fulcrum than the heavier person.
    c.   Closer to the fulcrum than the heavier person.
    d.   On the same side as the heavier person.
    e.   On the same side, but farther away than the heavier person.

Second class lever:

A wheelbarrow is a good example of a second class lever.

Here, the load in the wheelbarrow is effort balanced between the wheel (fulcrum) and the force (handlebars).

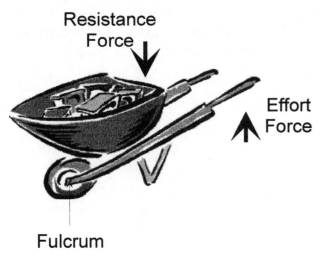

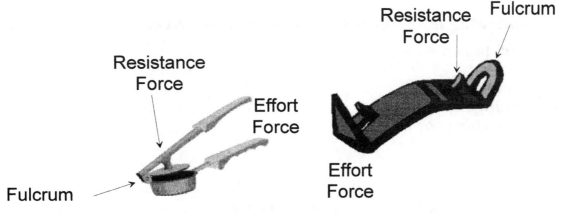

Other examples are the garlic press and the bottle opener.

Question 7:

Where is the resistance force in a second class lever?

    a.   Between the effort force and the fulcrum.

    b.   Farther out from the fulcrum than the effort force on the same side.

    c.   Farther out from the fulcrum than the effort force on the opposite side.

    d.   Closer to the fulcrum on the opposite side than the effort force.

    e.   None of these.

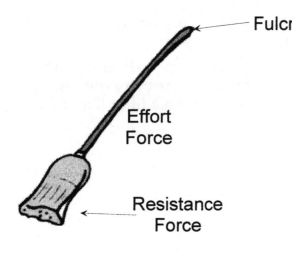

In a third class lever, the effort force is between the resistance force and the fulcrum. The most common example of a third class lever is a broom.

Question 8:

Which of the following is an example of a third class lever?

    a. a pull tab on a soda can

    b. a seesaw

    c. a bowling ball

    d. a rolling pin

    e. a revolving door

SECTION 3: MACHINES

Machines simplify work to be done. For example, imagine trying to open a can of corn without using a can opener.

The simple machines you should know are: lever, pulley, wedge, inclined plane, screw, and wheel & axle. Some of these machines are related to others, but are used in a different way. Combinations of these machines are called compound machines. (Scissors, wheelbarrows, and cranes are all examples of compound machines.)

Machines usually increase the distance through which a force must act in order to get work done, or they change the direction the force may act.

Work = force x distance.

Therefore if you increase the distance, you decrease the amount of force you must apply.

EXAMPLE:

Applying ten force units through five distance units equals fifty work units. If the distance is increased to ten, the amount of force needed to do the same amount of work would only be five units.

work = force x distance

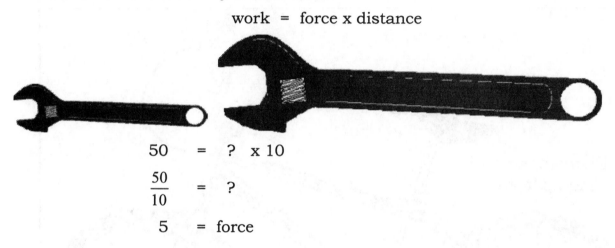

$$50 = ? \times 10$$

$$\frac{50}{10} = ?$$

$$5 = force$$

Question 9:

If you triple the force through the same distance, how much work would you do?

a. $\frac{1}{3}$

b. $\frac{1}{2}$

c. 1

d. 2

e. 3

SECTION 4: THE PULLEY

A pulley is a bearing to which a belt is attached. It can be used to change the direction of the belt, or to turn a shaft.

Points to remember:

1) All pulleys on the same shaft turn in the same direction.

2) All pulleys on the same shaft turn at the same speed.

Both pulleys on the shaft will make the same number of revolutions in a given amout of time.

Shaft

3) Different sized pulleys attached to different shafts by a "belt" (or other drive device) turn at different speeds. Comparison of speed is the inverse of the ratio of their diameter size.

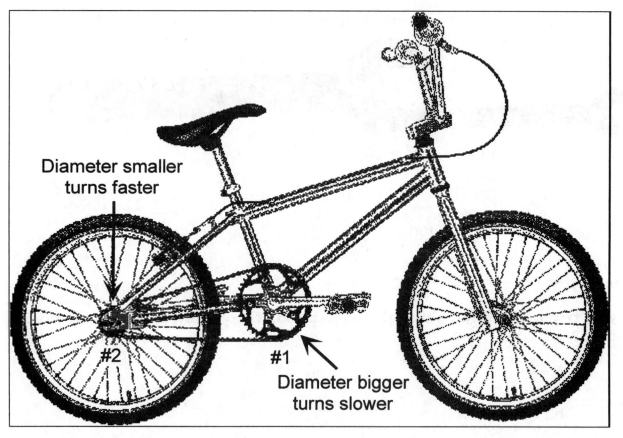

Diameter smaller turns faster

#2

#1

Diameter bigger turns slower

#2 pulley is half the diameter of pulley #1. The inverse is $\frac{2}{1}$ or 2. Therefore pulley #2 turns two times for every turn pulley #1 makes. The speed is twice as fast.

Question 10:

Which of the following is true of pulleys?

      a. different sized pulleys attached by a belt turn at different speeds
      b. pulleys on the same shaft turn at the same speed
      c. pulleys on the same shaft turn the same direction
      d. a and c only
      e. all of the above

The main purpose of creating pulley systems is to:

1)   Change the direction of the force.

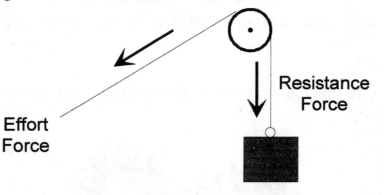

2)   Decrease the effort force needed to lift objects at the expense of adding distance through which the effort force must act.

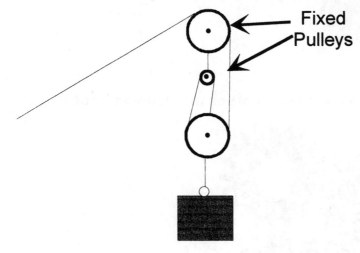

The greater the number of ropes connecting the pulleys, the easier it is to lift the resistance force.

Complex pulley systems are used on sailboats, cranes, and elevators. A very simple pulley arrangement is used on clothes dryers and car fans.

159

Question 11:

Which of the following does not have a pulley system?

    a. a clothes dryer
    b. a car fan
    c. a can opener
    d. an elevator
    e. a sailboat

SECTION 5: GEARS

Gears are similar to pulleys except that they have teeth to engage other gears instead of using belts.

Points to remember:

1)   One gear engaging another gear causes it to turn in the opposite direction.

2)   An odd number of gears engaging each other in a loop will not turn.

3) Different sized gears turn at different speeds. Comparison of speed is the inverse of the ratio of the number of teeth (cogs) on each gear.

The small gear makes 15 revolutions for every 8 revolutions of the big gear.

8 Teeth

15 Teeth

Question 12:

How would the seventh gear in a straight line series turn?

    a. In the same direction as the first gear.
    b. In the opposite direction as the first gear.
    c. Faster than the first gear.
    d. Slower than the first gear.
    e. There is not enough information given.

Cars use gears in their transmissions to change the amount of output work the motor produces and the number of times the wheels of the car turn.

Another example of gears is on a multi-speed bicycle. The gears are connected by a chain. By changing the gear ratio, the cyclist can bike up a hill using lots less effort than he/she would if using a bicycle without a gear system.

Question 13:

Why does the bicyclist change gears when going up a hill?

    a. to do more work
    b. to do less work
    c. to increase the effort force
    d. to decrease the effort force
    e. none of these

SECTION 6: INCLINED PLANES

A wedge is a simple machine that employs a large effort force to act through a very short distance. An ax, maul, chisel, nail, and knife are all wedges.

Question 14:

What is the purpose of a wedge?

      a. to increase the effort force
      b. to increase the resistance force
      c. to concentrate the effort force
      d. to absorb the effort force
      e. to multiply the effort force

Inclined Plane

It takes less effort force to move a resistance force up a ramp (inclined plane) than it would to lift that same force straight up to the desired height.

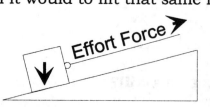

Resistance
Force

Adding distance allows the use of less effort to accomplish the same work amount of work.

EXAMPLE: Roads (going over hills) and ramps for wheelchairs.

Question 15:

Which of the following is not an inclined plane?

    a. a wheelchair ramp
    b. a ramp in a parking garage
    c. a gang plank on a ship
    d. an escape chute on an airplane
    e. an elevator in a hotel

A screw is a curved incline plane. The incline wraps around the central post.

Other examples are drill bits, light bulbs, and clamps.

Question 16:

Which of the following is not a curved incline plane?

    a. a screw
    b. a twist off bottle cap
    c. a driveway
    d. a spiraling parking garage ramp
    e. a drill bit

SECTION 7: WHEEL AND AXLE

Just remember this, a wheel and axle <u>reduces</u> friction. In many instances a lot of energy is wasted overcoming friction. Any wheel on an axle reduces the energy wasted. This means more work can be done for the same amount of effort. A machine employing wheel and axle mechanisms is more efficient.

Question 17:

What is the purpose of a wheel and axle?

    a. to make a car go faster
    b. to reduce friction
    c. to increase the resistance force
    d. to waste energy
    e. to increase friction

SECTION 8: HEAT AND ENERGY TRANSFER

Heat is the total energy of motion of all the molecules in an object. A measure of the speed the molecules are moving is called temperature.

Heat is transferred from areas of greater concentrations to areas of lesser concentrations - that is, from where it is hot to where it is cool. A melting ice cube in your hand is warming from the heat of your body.

Question 18:

Which of the following is not the way heat would flow?

      a. from ice to water
      b. from fire to water
      c. from your body to ice
      d. from boiling water to you
      e. from a hot frying pan to an egg

Heat may travel from one place to another by the following ways:

1) **conduction** - direct contact between the heat source and the object.

2) **convection** - "heat rises" = actually air or water come in contact with a heated object and rise because the heat makes the air or water less dense than surrounding unheated air or water. Thus, the heated fluid carries the heat from one place to another.

3) **radiation** - here, heat travels through any medium by wave motion (like the heat of the sun).

Heat travels best through conductors. Good conductors are usually metals - gold, silver, copper, aluminum, and steel are conductors of heat energy. Gold, silver and copper are perhaps the best.

Anything that does not conduct heat well is said to be a non-conductor, or insulator.

Question 19:

Which of the following is not a good conductor?

      a. copper
      b. silver
      c. gold
      d. wood
      e. aluminum

SECTION 9: REFLECTION AND REFRACTION

When you look in a mirror you see your reflection. Reflection is the bouncing back of energy waves off a surface. Sunlight reflects off glass, the ocean, and even the moon.

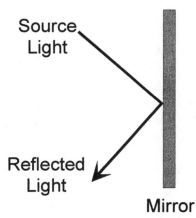

Light reflects off a flat mirror at the same angle it strikes that mirror.

Question 20:

Which of the following does light bounce off of?

      a. mirrors
      b. glass
      c. water
      d. the moon
      e. all of the above

Another event that may occur when a form of wave energy enters into a different medium (example: light entering water), is refraction. Refraction may be described as the bending of the energy wave.

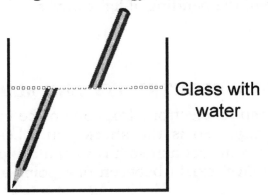

Glass with water

The pencil is not broken. What has happened here is that light waves have been bent (refracted) because they move faster upon entering the water.

Question 21:

What happens to light when it enters a different medium?

        a. it glows
        b. it breaks
        c. it bends
        d. it twists
        e. none of these

Mirage

Air

Real Tree

A mirage is an example of the bending of light waves.

SECTION 10: ELECTRICITY AND MAGNETISM

Electricity is the movement of electrons from one place to another. Lightning is a form of static electricity. So is the shock you receive when you touch a doorknob after dragging your feet across a rug on a dry day. As long as a large enough potential (difference) exists between one point and another, electricity will flow.

Electricity flows best through a conductor. The best conductors are gold, silver and copper. Things that do not conduct electricity well are called insulators. Glass, plastic, ceramic, and wood are all insulators.

Question 22:

Which of the following are good insulators?

    a. copper
    b. wood
    c. silver
    d. a and c only
    e. all of the above

The amount of electricity that flows through a wire is known as the current. The current is measured in units called **amperes** or **amps**.

The electrical push through the wires is the electromotive force called **voltage** and is measured in **volts**.

There is always resistance to the flow of electricity. This is called **resistance** and is measured in **ohms** (Ω).

Question 23:

Match the word in column A with the measurement in B.

          <u>A</u>   <u>B</u>

    1. resistance   a. volts

    2. current   b. amps

    3. voltage   c. ohms

The relationship of the current (I), voltage (E), and resistance (R) is shown in Ohm's Law:

$$\text{Voltage (E)} = \text{current (I)} \times \text{Resistance (R)} \quad E = IR$$

A good way of remembering this formula is to use the following diagram:

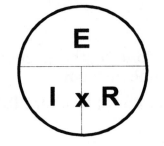

In order to solve problems using Ohm's Law, just cover what you don't know and solve. For example: How much voltage is required to provide 50 Amps passing through 10Ω resistance?

50 x 10 = 500 volts

EXAMPLE: What is the resistance of a T.V. set if the voltage is 120V and the current is 6 amps?

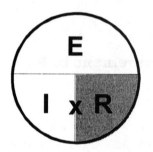

$$\frac{E}{I} = \frac{120V}{6\,amps} = 20\Omega$$

Question 24:

What is the resistance if the current is 60 amps and the voltage is 120 volts?

    a. 1/2 ohm
    b. 2 ohms
    c. 30 ohms
    d. 60 ohms
    e. 7200 ohms

Power is the amount of work done by the electricity. It is measured in **watts**. We have the same kind of formula as above:

Power (P) = current (I) x voltage (E) or, P = I x E.

Solve any problems as you would using Ohm's Law.

Question 25:

What is the power if the current is ten amps and the voltage is 120 volts?

     a. 1/12
     b. 1.2
     c. 12
     d. 120
     e. 1200

Electricity flows through a circuit unless we're dealing with static electricity. A circuit is an unbroken path in which electricity flows to and from the source.

There are two types of circuits: **Series** and **Parallel**.

Series Circuit

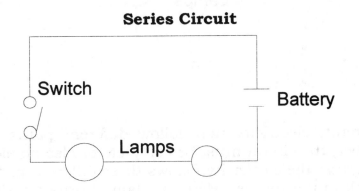

Pictured above is an open circuit. No electricity will flow until the path is completed by closing the switch.

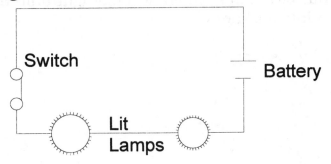

If we were to disconnect a lamp, the circuit would be broken and the other lamp would go out.

Electricity entering our homes must pass through a fuse box or circuit breaker box. This is wired in a series to protect us from electrical overloads, which can cause a fire. Lights on a single switch are also series wired. Any appliance operated with a switch is series wired. However, the circuit wiring in a house is parallel, not series.

Question 26:

Why is the circuit breaker wired in series?

  a. That is just the way it is done.
  b. To protect against overloads.
  c. To make the lights flash.
  d. To save energy.
  e. So switches can be run.

Parallel Circuits

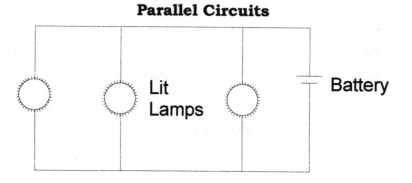

In a parallel circuit, electricity may follow different paths to return to the source. Remember, the circuit must be complete or else no electricity will flow. In the above diagram, the electricity follows three different paths: through each lamp (3) and back to the source. Here, if a lamp burns out disconnecting that part of the circuit, the electricity continues to flow through the other two lamps back to the source.

This is similar to your own home, where when a light bulb burns out the entire house is not thrown into darkness.

Question 27:

Why are some circuts run in parallel?

        a.   to keep lights from burning out
        b.   to prevent circut overloads
        c.   so you don't lose all your power if one electrical appliance burns out
        d.   to save energy
        e.   none of these

SECTION 11: MAGNETISM

A natural magnet is called a lodestone. It is a rock that has a high percentage of iron (called magnetite) in its chemical make-up.

Magnets have two poles: north and south. Emanating from these poles are invisible magnetic lines of force. If we sprinkle iron filings on top of a piece of paper covering a magnet, the iron filings line up on the lines of force.

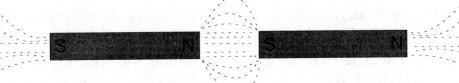

Notice, N and S attract. This conforms to opposites attract and likes repel. Two N/N or S/S would show a pattern like.

Question 28:

What are the poles of a magnet called?

        a. east and west
        b. east and north
        c. west and north
        d. east and south
        e. north and south

Magnetism and electricity are related. If we were to pass a magnet through a coil of wire connected to a galvanometer, the needle would move as long as the magnet were in motion.

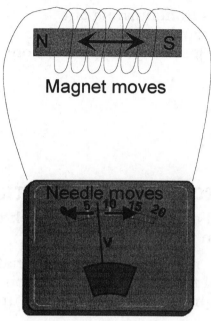

Magnet moves

Needle moves

What is happening is the magnetic lines of force are causing electrons to move back and forth in the wire (alternating current).

This "induction" effect is used to generate electricity. Atomic energy, running water (rivers, waterfalls, dams), gas or coal fuel provide the power to turn large turbines which have huge magnets and wire coils turning in a field of magnetic lines of force. This generates huge amounts of electricity, which eventually is divided and dealt out to all the electric company's consumers.

Question 29:

What is the induction effect?

  a. An alternating current.
  b. Wire coils and magnets.
  c. The burning of coal.
  d. The flow of electrons due to a magnetic field.
  e. None of these.

SECTION 12: FISSION AND FUSION

Fission

Atomic energy has been with us, in usable form, for the past fifty years. This type of energy is best understood as the **radioactive decay process**.

Unstable atoms are always undergoing the decay process. In a given amount of time, half of the radioactive substance will have decayed into daughter elements. Scientists can determine the half-life of radioactive substances.

EXAMPLE: U^{238} (Uranium with an atomic weight of 238) has a half-life of 4,500,000,000 years (4.5 billion years). This means that one pound of U^{238} existing 4.5 billion years ago is only a half pound of U^{238} today. The other half pound is now converted into **daughter elements** such as Radon (Rn), Thorium (Th), Radium (Ra), and Lead (Pb).

$$X = U^{238} \text{ atoms}$$

$$Y = \text{decayed } U^{238} \text{ atoms}$$

1 lb. U^{238}
4.5 billion years ago

Now 1/2 lb. U^{238}
1/2 lb. daughter elements

Question 30:

Into which of the following does uranium not decay?

    a. thorium
    b. radon
    c. radium
    d. lead
    e. none of these

In order to calculate the amount of original radioactive elements remaining after a number of half-lives, one must use the formula:

$$\left(\frac{1}{2}\right)^n$$

EXAMPLE:

A substance passed through four half-lives. How much of the original substance remains? How much has decayed?

$$\left(\frac{1}{2}\right)^n = \left(\frac{1}{2}\right)^4 = \left(\frac{1}{2}\right)\left(\frac{1}{2}\right)\left(\frac{1}{2}\right)\left(\frac{1}{2}\right) = \frac{1}{16}$$

Only $\frac{1}{16}$ of the original radioactive substance remains, and $\frac{15}{16}$ has decayed.

$$\frac{1}{16} + \frac{15}{16} = \frac{16}{16} = 1$$

EXAMPLE:

A radioactive substance has a half-life of twenty-five years and passes through five half-lives.

  a) How many years does it take to pass through five half-lives?

  b) How much of the original material remains?

  c) How many years will it be before 99% of the original substance decays?

  a) $\left(\dfrac{25 \text{ years}}{\text{half life}}\right)(5 \text{ half lives}) = 125 \text{ years}$

  b) $\left(\dfrac{1}{2}\right)^n = \left(\dfrac{1}{2}\right)^5 = \left(\dfrac{1}{2}\right)\left(\dfrac{1}{2}\right)\left(\dfrac{1}{2}\right)\left(\dfrac{1}{2}\right)\left(\dfrac{1}{2}\right) = \dfrac{1}{32}$ of the original substance remains.

  (c) $\left(\dfrac{1}{2}\right)^n = \left(\dfrac{1}{2}\right)^2 = \dfrac{1}{4}$

   $\left(\dfrac{1}{2}\right)^n = \left(\dfrac{1}{2}\right)^3 = \dfrac{1}{8}$

   $\left(\dfrac{1}{2}\right)^n = \left(\dfrac{1}{2}\right)^4 = \dfrac{1}{16}$

   $\left(\dfrac{1}{2}\right)^n = \left(\dfrac{1}{2}\right)^5 = \dfrac{1}{32}$

   $\left(\dfrac{1}{2}\right)^n = \left(\dfrac{1}{2}\right)^6 = \dfrac{1}{64}$

   $\left(\dfrac{1}{2}\right)^n = \left(\dfrac{1}{2}\right)^7 = \dfrac{1}{128}$

After seven half-lives less than 1% of the original substance remains.

$$7 \times 25 \text{ years} = 175 \text{ years}$$

Question 31:

How much of the original material is left if the material has a half-life of twenty years and the time elapsed is 100 years?

a. $\dfrac{1}{64}$

b. $\dfrac{1}{32}$

c. $\dfrac{1}{16}$

d. $\dfrac{31}{32}$

e. $\dfrac{63}{64}$

The reason for decay is instability. This probably has something to do with the amount of neutrons in the nucleus of the atom; protons do not seem to decay.

Two kinds of particles are thrown off during decay:

Alpha (α) and Beta (B)

An alpha particle is comparable to a helium nucleus:

$$\alpha = {}_{2}^{4}\text{He}$$

A beta particle is comparable to an electron. Now I know you're saying that electrons don't exist in the nucleus - true! Scientists believe a neutron decays (breaks-up) into a proton (+), electron (-), and a neutrino. ($n^0 = p^+ + e^- +$ neutrino). The neutrino goes somewhere. It's a very elusive particle. The electron leaves the nucleus, and the proton stays behind.

Question 32:

What is a beta particle like?

a. an electron
b. a proton
c. two protons and two neutrons
d. gamma rays
e. two protons

Also leaving the nucleus during decay are gamma rays (γ).

During the decay of U^{238}, both α and B particles are thrown off until a stable form of Pb is produced.

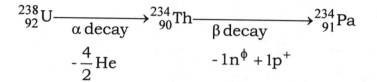

α decay = loss of 2 p$^+$ (at. no. [92 to 90]) and

loss of 4 particles (2p$^+$ & 2n$^\phi$) from atomic weight.

B Decay = loss of 1 neutron, gain of 1 proton

= no loss at. wt., gain of 1 p$^+$ = gain in at. no. (90 to 91).

Decay of ^{238}U

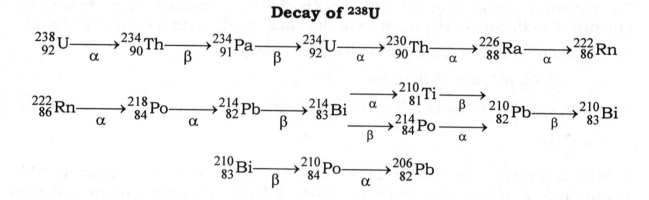

Fusion

Whereas fission is the decay of elements, fusion is the building of elements. Scientists believe the sun produces its energy by the fusion process. Simply, that hydrogen nuclei combine to form helium nuclei.

$$4 \text{ H} \rightarrow \text{He}$$

The energy produced in this reaction is tremendous - more than fission. Einstein's equation, $E = MC^2$ explains the energy production.

Remember, the sun consumes four tons of H per second.

Now applying numbers to illustrate energy production:

$E = MC^2$

energy = matter "lost" x (speed of light)

ENERGY = 4 tons x (186,000 mi/sec)^2

large number = 4 x 186,000 x 186,000

The reaction is:

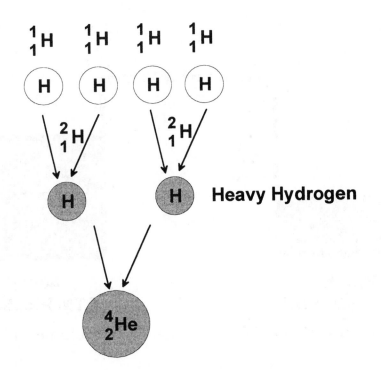

Heavy Hydrogen

Question 33:

Which of the following is true?

      a. Fusion creates matter.
      b. Fusion releases more energy than fission.
      c. Coal burning produces more energy than fission.
      d. Fission products are always radioactive.
      e. All of the above.

SECTION 13: GAS LAWS

Charles' Law

The volume of a gas is directly proportional to its temperature (when the pressure remains constant).

Temperature & Volume

200° K
1 ATM PRESSURE

400° K
1 ATM PRESSURE

If the Temperature rises from 200° K to 400° K , the volume will also increase.

Question 34:

In the example above how much would the volume increase?

    a. 50%
    b. 100%
    c. 150%
    d. 200%
    e. 400%

Boyle's Law

Volume of a gas varies inversely with its pressure.

$$p \, \alpha \, \frac{1}{V}$$

As pressure increases, volume decreases.

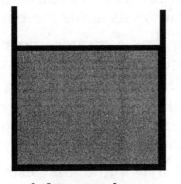

1 Atmosphere   **2 Atmospheres**

$$V = \frac{1}{2}$$

Question 35:

Which of the following is not true?

    a. As pressure increases, volume decreases.
    b. The volume at a given pressure only depends on the pressure.
    c. Pressure and volume are dependent on each other.
    d. a and b only.
    e. All of the above.

1. d. The flight path of a bullet would be similar to that of a cannon. The last paragraph says that the path of a cannon is curved because of the friction of the air and the force of gravity pulls it down.

2. c. This is the first case, where an object at rest stays at rest. The marble stays where it is put the same way that the book stayed where it was placed.

3. b. From the paragraphs, we know that F = M x A. Since we are given the force and the mass, if we divide the force by the mass we will get acceleration. Nine divided by three equals three.

4. e. When the ball is thrown, the third law of motion states that the boat must move also in the opposite direction. Because the boat is so much more massive, it would not move as far as the baseball.

5. a. The question states that the moon has a greater impact on the tides than the sun due to its gravitational pull. This means that the moon has a greater attraction to the Earth than the sun.

6. b. The example of the seesaw says that a lighter person can balance a heavier person if the heavier person is sitting closer to the fulcrum. This means the lighter person must be sitting farther away.

7. a. The example of the wheelbarrow shows that the resistance force is between the effort force and the fulcrum.

8. e. The best example of a third class lever given is the revolving door. You push on the door in the middle of a side. The resistance is at the edge of a door and the fulcrum is in the center.

9. e. The example says that work is equal to force times distance. If the distance is the same and the force is tripled, the work must also be tripled.

10. e. All of the statements about pulleys are true. Different sized pulleys attached to different shafts turn at different speeds. Pulleys on the same shaft rotate at the same speed and turn in the same direction.

11. c. A can opener does not use a pulley system because a pulley system cannot generate enough force in a small area. It uses a gear system instead. The clothes dryer and car fan are given as examples of simple pulley systems. The elevator and sailboat are complex systems.

12. a. No information is given on the size of the gears so there is no way to say how fast any of the gears are turning. The paragraph does say that touching gears turn in opposite directions. The second gear

turns opposite the first. The third turns opposite the second, or the same way as the first. The fourth turns opposite the first and the fifth the same as the first. The sixth turns the opposite of the first. Finally, the seventh is the same as the first.

13. d. The amount of work being done would remain the same. Shifting gears allows the bicyclist to reduce the amount of effort force while he is going up the hill.

14. c. The paragraph says that a wedge takes a large effort force and makes it act through a small distance. This is the same as saying that a wedge is used to concentrate your force to a specific spot.

15. e. The elevator is the only choice that goes straight up and down, and is therefore not an inclined plane. An escape chute and a gang plank both slope down to the ground. A wheelchair ramp and a parking garage ramp are both sloped upwards.

16. c. The screw and the drill bit are given as examples of curved inclined planes. A twist off bottle cap has curved grooves to make it a curved incline plane. The spiraling parking garage ramp could also be a curved incline plane. A straight driveway does not fit the curved inclined plane definition.

17. b. The main focus of the paragraph is a wheel and axle reduces friction.

18. a. The last paragraph states that heat flows from areas of high concentrations to areas of low concentrations. The only choice that has heat going from cool to hot is from ice to water.

19. d. Gold, silver, copper, and aluminum are all listed as good conductors. That leaves you with wood which is a poor conductor.

20. e. The paragraphs state light bounces off all of the choices given.

21. c. The paragraph states light bends when it enters the water and that causes the pencil to look bent.

22. d. The paragraph states that gold, silver, and copper are all good insulators. Therefore answers c, silver, and a, copper, are correct. Answer b, wood is an insulator. Therefore answer d, both a and c, is correct.

23. 1-c

2-b

3-a

24. b. $\dfrac{E}{I \times R}$ $R = \dfrac{E}{I}$ $R = \dfrac{120}{60}$ $R = 2$

25. e. $P = I \times E = 10 \times 120 = 1200$

181

26. b. The paragraph states the circuit breaker is wired in series to protect us against overloads.

27. c. The section on parallel circuits said if one section of the circuit broke, the other sections would still function. This is why circuits are run in parallel.

28. e. The last paragraph states the poles of a magnet are called the north and south.

29. d. The induction effect is the flow of electrons due to movement in a magnetic field. Wire coils, magnets and coal are used to produce the effect. Alternating current can be a result of the effect.

30. e. All four of the choices were listed as daughter elements from the decaying of uranium.

31. b. In 100 years the material would go through five half lives.

    $100/20 = 5$

    $(1/2) \times (1/2) = 1/4$ 2 lives

    $(1/4) \times (1/2) = 1/8$ 3 lives

    $(1/8) \times (1/2) = 1/16$ 4 lives

    $(1/16) \times (1/2) = 1/32$ 5 lives

32. a. The paragraph states a beta particle is the same as an electron.

33. b. The first paragraph states fusion releases more energy than fission.

34. b. In the example, the temperature doubles. You are also told that volume increases at the same rate temperature increases. Therefore it would double also. In order for the volume to double it must increase 100%.

35. b. Taking the choices one at a time, we have:

    a. As pressure increases, the volume decreases. This is a true statement. Things get smaller when placed under pressure.

    b. The volume at a given pressure depends only on the pressure. What this is saying is if you know the pressure, you also know the volume. This is false. You also need to know the temperature of the object.

    c. Pressure and volume are dependent on each other. What this is saying is if you change one, the other one will also change. This is a true statement.

    Therefore, only b is a false answer.

As you take this short quiz, remember that the information you need to answer the question is presented in the paragraph, chart, or diagram. You must simply analyze the information properly.

Physics Chapter Test

Question 1:

The second law of motion states that force is equal to mass times acceleration. How much more force would be required to accelerate a fifty-ton ship at the same rate as a ten-ton ship?

    a. The same
    b. Two times
    c. Five times
    d. Ten times
    e. Fifty times

Question 2:

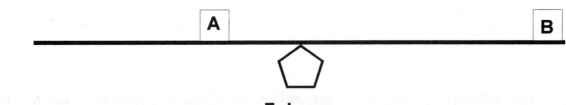

Fulcrum

Work is equal to force times the distance. This means that to do the same amount of work, a greater distance requires less force. Which block in this diagram weighs more?

    a. Block A
    b. Block B
    c. They both weigh the same
    d. It depends on how strong gravity is.
    e. There is not enough information given.

Question 3:

Pulleys are used in two cases. The first instance where pulleys are used is to change the direction of a force. If you were trying to raise a bale of hay to the loft in a barn, it would be impractical to climb on the roof and pull the bale straight up. Instead, it is much easier to run the rope through a pulley above the loft. Then you can pull the bale up while standing on the ground.

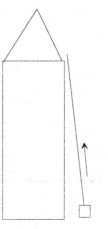 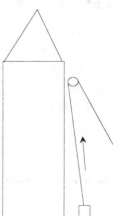

Which of the following uses a single pulley to change the direction of the force?

    a. A flagpole
    b. A chainsaw
    c. An automobile engine
    d. A diving board
    e. An elevator

Question 4:

The second way pulleys are used is to increase the distance through which a force acts. This means that more work can be done for the same amount of force. At least two pulleys must be used, and one pulley must be able to move.

Diagram A

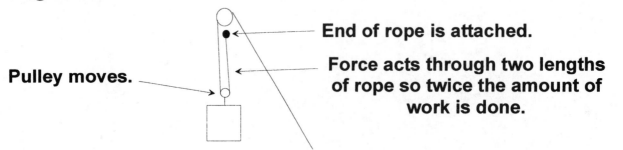

End of rope is attached.

Force acts through two lengths of rope so twice the amount of work is done.

Pulley moves.

Diagram B

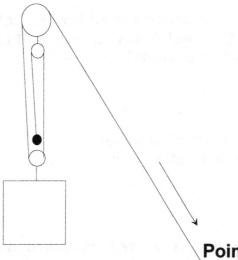

Point A

If the block in diagram B weighs six hundred pounds, how much force must be applied at point A to lift it?

    a. 1800 pounds
    b. 600 pounds
    c. 300 pounds
    d. 200 pounds
    e. 100 pounds

Question 5:

The first law of motion states a body at rest stays at rest and a body in motion stays in motion. Which of the following is not an example of this law?

    a. A marble on a flat desk remains in one place.
    b. A marble on a flat desk gets bumped and slowly rolls completely off the desk.
    c. A car traveling down the road continues to move forward even after the foot is taken off the accelerator
    d. A boat stays in the center of the lake without moving for two days.
    e. None of the above

Question 6:

Ohm's law states voltage = current x resistance. This is more commonly written as E = I x R. If the resistance is cut in half and the voltage is doubled, what happens to the current?

  a. It doubles.
  b. It is cut in half.
  c. It quadruples.
  d. It is cut to on fourth the original.
  e. There is not enough information given.

Question 7:

The volume of a gas decreases as the pressure on it increases. What happens to the volume of a scuba tank as a diver dives into deeper water?

  a. The volume stays the same.
  b. The volume gets bigger.
  c. The volume gets smaller.
  d. It depends on whether the diver is diving in the ocean or a lake.
  e. There is not enough information given.

Question 8:

The ratio by which a set of gears spins is inversely proportional to the ratio of the number of teeth on each gear.

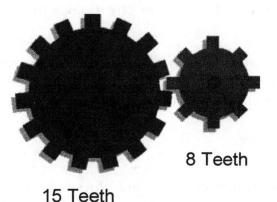

8 Teeth

15 Teeth

If the larger gear makes one hundred and twenty revolutions, how many revolutions will the smaller gear make?

     a. 120
     b. 240
     c. 64
     d. 225
     e. It depends on the size of the gears.

Question 9:

The force, due to gravity, depends on the mass of the two objects and the distance they are separated. Which of the following is true about the attraction between a person and the Earth?

     a. It increases as he moves up and away from the surface.
     b. It decreases as he moves up and away from the surface.
     c. It increases as he moves towards the equator.
     d. It decreases as he moves below the surface.
     e. It remains the same regardless of where you go.

Physics Chapter Questions

1. c. Force equals mass times acceleration. The new mass is five times greater than the old mass. Therefore, the force must be five times as great to have the same acceleration.

2. a. The two blocks are in balance. Therefore, the block closer to the pivot is the heavier block.

3. a. A pulley is placed at the top of a flagpole. This allows the flag to be easily pulled to the top of the pole while the puller stays on the ground.

4. d. There are three lengths of rope supporting the block. Therefore, the force needed to lift the block would be one-third of its weight. One-third times six hundred pounds equals two hundred pounds.

$$\frac{1}{3} \times 600 = 200$$

5. e. All are examples of this law. Answers a and d are examples of staying at rest. Answers b and c are examples of bodies staying in motion.

6. c. Two times the old voltage divided by one half the old resistance would give you four times the old current.

7. a. A scuba tank is rigid, so it would not change with the increase in pressure. If it had been a balloon, it would have gotten smaller as the diver went down deeper.

8. d. The smaller gear makes fifteen rotations for every eight the large gear makes. One hundred and twenty divided by eight equals fifteen. Fifteen times fifteen equals two hundred and twenty-five.
$$\frac{120}{8} = 15 \qquad 15 \times 15 = 225$$

9. b. Climbing up on the surface of the Earth means the person is moving farther from the center of the Earth. The farther two objects are separated, the less their gravitational attraction.

Scientific Method

The last thing we need to talk about is something called the **scientific method**. It is a procedure followed in most experiments and is designed to give results that are consistent and valid. This means after someone does an experiment and gets results, another person will be able to repeat the experiment and get the same results.

Hypothesis

The first step in the scientific method is coming up with a hypothesis. This is the scientist's best guess as to how things will work. Once a hypothesis is formed, an experiment is designed to prove or disprove the hypothesis.

Experiment

The experiment must contain certain things. First of all you need a complete list of the quantities of everything you will use. In each experiment, you also need a set **procedure** (how the experiment will be carried out) and a **control** (a group that remains the same and is compared to the changed group). The control group is very important. If you do not maintain a control you will have no idea if the results you observe are a result of the changes you intentionally made, or due to other random factors beyond your control.

Procedure

The procedure must be very clear. It not only states what is to be done. It states their order in a series of steps.

EXAMPLE:

Step 1: Take one liter of water

Step 2: Heat to boiling

Step 3: Add to Calcium Carbonate

The Control

The control of the experiment is extremely important. The results of an experiment can never be truly valid unless there is a control group. For example: If you were testing a drug on mice to see if they had more energy and you did not have a group that did not get the drug, you would have nothing with which to compare your results.

Observation

Here you observe your experiment as it is taking place and keep track of your results. You must keep detailed records of not only what is occurring in your experiment, but also anything that could affect your experiment. For example, if you are growing plants, is it sunny or cloudy? This is important because when someone tries to recreate your experiment and does not get the same results, they can then check some of these outside factors to see if those factors caused the differing results.

Results

Here you list the results of the experiment. This is simply a statement of what happened in your observation.

Conclusion

After you do your experiment, you must look at your results and relate it back to your original hypothesis. The results can either confirm your hypothesis, or they can disprove your hypothesis. If they disprove your hypothesis it does not mean that the experiment was a failure. Often it is more important to know what doesn't work and why than it is to know what does work, but not know why.

Many breakthroughs in science are made by looking at results that did not fit the original hypothesis. A new hypothesis is formed based on those results and then more experiments are performed to see if the new hypothesis is indeed correct.

This completes our study for the GED Science Test. You are now ready to take the sample science test which follows. Each question or group of questions will have a reading passage. You will be able to answer the questions based on what is stated or inferred in the passage, graph or diagram.

Chemical Name	Chemical Symbol
Calcium	Ca
Chlorine	Cl
Magnesium	Mg
Oxygen	O
Potassium	K
Sodium	Na
Sulphur	S

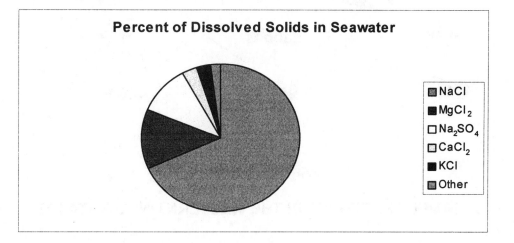

1. **Using the graph above, approximately what percentage of the dissolved solids in seawater is sodium chloride?**

    a.   10%
    b.   12%
    c.   35%
    d.   50%
    e.   65%

2. **Freight cars are being assembled into a train. One boxcar rolls down the track into one that is standing still. The two cars couple and then they continue to roll together. Their speed is now less than the speed of the original boxcar by itself. This is an example of the Law of Conservation of Momentum. Which of the following would be the best example of this law in the real world?**

    a.   a car sliding out of control in a sharp curve
    b.   a truck pulling two trailers
    c.   pool balls striking each other on a pool table
    d.   a jack hammer breaking concrete
    e.   an air plane suspended in the air by air pressure

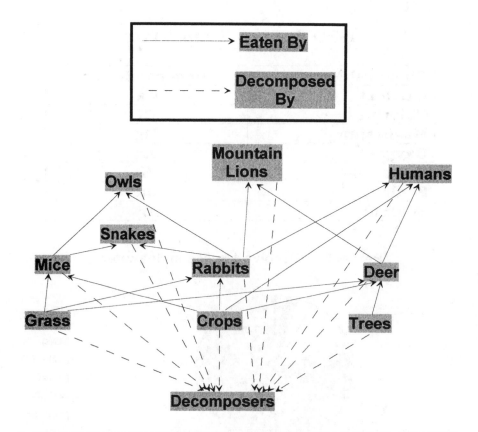

A SAMPLE FOOD WEB IN THE WESTERN UNITED STATES

A food web is a diagram of the interrelationships among organisms that live in a given community. In this diagram, the food animals eat is represented by the solid arrows. The head of the arrow points to the animal doing the eating. The dashed arrows trace the paths of the decomposers.

Autotrophs – organisms that produce their own food from sunlight
Herbivores – animals that feed only on plants
Carnivores – animals that only eat animals
Omnivores – animals that eat both plants and animals
Decomposers – breakdown wastes, dead plants and dead animals

3. **Based on this food web, which type of organism would a mountain lion be classified as?**

      a.   an autotroph
      b.   a herbivore
      c.   a decomposer
      d.   a carnivore
      e.   an omnivore

4. Rabbits and deer feed on which of the following?

    a.   grass and mice
    b.   autotrophs
    c.   decomposers and crops
    d.   herbivores
    e.   trees

5. The "Plate Tectonic Theory" holds that the Earth's surface is made of plates that move on a partially melted layer of the mantle. This layer is called the asthenosphere. It remains melted due the heat given off by the radioactive decay of elements in the Earth.

What will cause the plates to eventually stop moving?

    a.   They will finally get where they belong.
    b.   The Earth's interior will cool.
    c.   The amount of energy reaching the Earth from the sun is expected to decrease.
    d.   There will be a huge increase in the number of clouds due to global warming from the "Greenhouse Effect"
    e.   The Earth's magnetic field will fail and the plates will stop moving.

6. The basic building block of all life is the cell. Most cells are similar in many ways, however there are some differences between plant and animal cells. First of all, since plants do no move on their own, their cells are surrounded by a rigid cell wall. The cell wall performs two main functions, support and protection. The second major difference is some plant cells contain chloroplasts. This is the site where photosynthesis, the conversion of sunlight's energy to useable energy, occurs.

If cell walls give plant cells a much higher degree of protection than the cell membranes that surround animal cells, why are animal cells not surrounded by a cell wall?

    a.   The rigidness of cell walls would make it difficult for animals to move.
    b.   Cell walls make it extremely difficult to intake food and since animal cells use so much more food they would die.
    c.   Cell walls do not allow water to pass, and animal cells would dehydrate if they had them.
    d.   Cell walls are toxic to animals.
    e.   Animal cells digest cell walls.

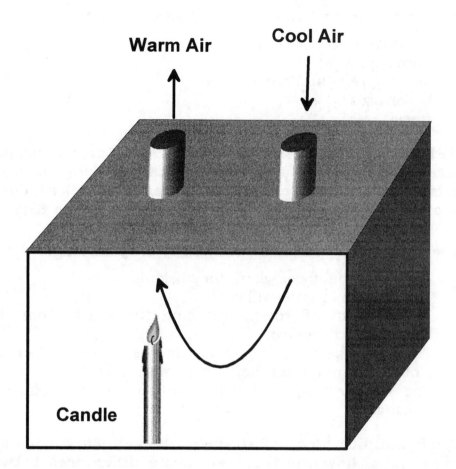

Warm Air

Cool Air

Candle

7. This is a simple convection cell. The warm air from the candle rises and goes out the closest pipe. This sucks cooler air in through the other pipe. The flow of air keeps the candle lit. If both vents are blocked, the candle goes out. Why?

    a.   Cold air is necessary to keep the candle from getting too hot.

    b.   The oxygen in the trapped air is exhausted.

    c.   Hot air must be able to escape if the candle is to continue burning.

    d.   A burning candle produces water and carbon dioxide. The water extinguishes the candle in a closed system.

    e.   A burning candle produces water and carbon dioxide. The carbon dioxide extinguishes the candle in a closed system.

8. Ocean water was not initially salty. Rivers pick up small amounts of minerals as they flow to the oceans. Water then evaporates from the oceans and forms rain, which then flows back along the rivers with new minerals to the ocean. Over time, this results in a build up of minerals in the ocean.

   Which of the following statements about rivers is true?
   a. Rivers get saltier over time.
   b. Rivers are responsible for vast deposits of salt that are mined and sold for seasoning food.
   c. When a river empties into the sea, it decreases the salinity of the nearby ocean.
   d. Rivers do no affect the salinity of water in the ocean.
   e. Ocean water and river water mix immediately.

9. A pot of cold water is placed on the stove. The burner is turned on and the temperature rises steadily until it reaches 100° C (212° F). The water begins to bubble vigorously, but the temperature does not change. The level of the water in the pan falls slowly.

   Which statement best describes what is happening?
   a. The water is absorbing all the energy from the burner. This energy is used to change from the liquid state to the gaseous state at the same temperature.
   b. The water is incapable of absorbing any more energy once it reaches the temperature of 100° C and the excess energy is absorbed into the air.
   c. The burner is probably not capable of supplying enough heat to raise the temperature above 100° C.
   d. Heat energy is wasted as it flows up the outside of the pot.
   e. The water is cooling the burner down. It keeps the burner from exceeding 100° C.

10. **Three five-hundred pound weights are being lifted to a height of six feet. Ten people are evenly spaced around the first weight and it is lifted straight up onto the platform. The second weight is lifted by a block-and-tackle pulley system, and the third weight is put on a cart and wheeled up a ramp.**

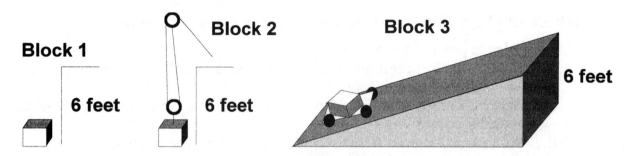

Which weight required the most work to raise it up to the platform?

   a.   Block number one required the most work because it took ten people to raise it.

   b.   Block number two took the most work because of all the rope needed to be pulled through the pulley system.

   c.   Blocks one and two took the same amount of work but block three took less because wheels were used to roll the block up the ramp.

   d.   All three blocks required the same amount of work.

   e.   There is not enough information given. It is necessary to know the shape and composition of each block.

11. **The moon's bright surface is always pointing directly towards the sun. It is bright because it is reflecting the sun's rays. Half of the moon is always bright, but most of the time some of the bright surface is pointing away from the earth, and we see the part of the moon that is in shadow.**

Why can a lunar eclipse only occur during a full moon?

   a.   At other times the sun's position changes too quickly.

   b.   At other times the moon is too far away to fall in the Earth's shadow.

   c.   A lunar eclipse can only occur at high tide.

   d.   The energy from the sun must be reduced during a lunar eclipse and this only occurs during a full moon.

   e.   The shadow from the Earth must fall on the moon, so the bright side of the moon must be facing towards the sun and the Earth.

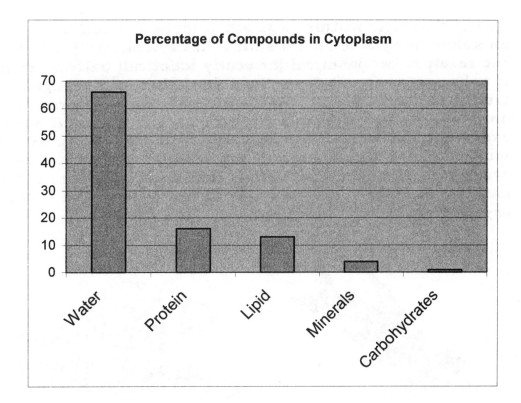

Percentage of Compounds in Cytoplasm

12. What is the most abundant compound in cytoplasm?

    a.   water
    b.   protein
    c.   lipids
    d.   minerals
    e.   carbohydrates

13. Cytoplasm contains two types of molecules, organic and inorganic. Organic molecules are made by living cells and include proteins, lipids (fats), and carbohydrates. Inorganic molecules are not made by cells and exist naturally in nature. They include water and minerals.

Which of the following are not made by cells?

    a.   water and proteins
    b.   proteins, lipids and carbohydrates
    c.   minerals and water
    d.   proteins and minerals
    e.   lipids and carbohydrates

14. One of the major features a scientist looks at when trying to classify an animal is symmetry. An animal has symmetry if it can be cut so the result is two identical (or nearly identical) halves. For instance, if a human were cut from the top of the head straight down the middle of the two legs, two identical pieces would result. There is only way to cut a human into two identical pieces. Humans are an example of an animal with bilateral symmetry. Other animals have multiple lines along which they may be cut. They have radial symmetry. Finally, the last group is animals which cannot be cut in half in any way to yield identical halves. They are termed asymmetrical.

Which of the following animals have bilateral symmetry?

    a.   1 and 4
    b.   5, 4 and2
    c.   5, 1 and 4
    d.   2 and 1
    e.   4, 1 and 2

15. An amoeba, an asymmetrical organism, has the same symmetry as which of these animals?

    a.   1
    b.   2
    c.   3
    d.   5
    e.   none of these

16. pH measures how strong an acid is. Every solution can be placed on the scale, which ranges from 1 to 14 (the lower the number the stronger the acid). Solutions with a pH of 7 are neutral. Solutions with a pH above 7 are basic.

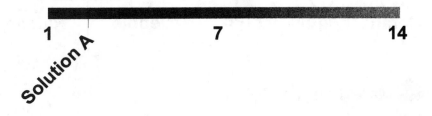

How would solution A be classified?

    a.   strong acid
    b.   weak acid
    c.   neutral
    d.   weak base
    e.   strong base

The ability to roll your tongue is controlled by a dominant gene in humans. If a child gets this gene from either parent then the child will be able to roll his tongue. In order for someone to not be able to roll their tongue, they must be homozygous (both copies of the gene the same) recessive. If a person has one of each type of gene, they are heterozygous.

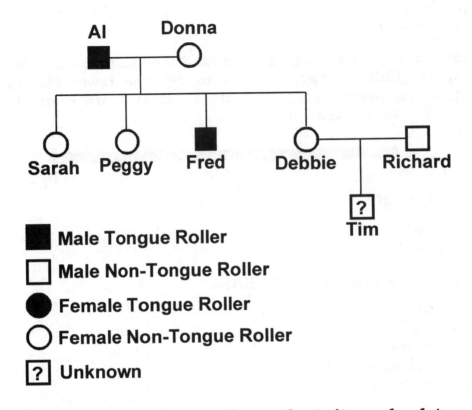

■ Male Tongue Roller

☐ Male Non-Tongue Roller

● Female Tongue Roller

○ Female Non-Tongue Roller

?︎ Unknown

17. From the passage and the pedigree chart, it can be determined that Tim is:

    a.   a homozygous tongue roller
    b.   a non-tongue roller
    c.   a heterozygous tongue roller
    d.   a heterozygous non-tongue roller
    e.   not enough information is given

18. Fred marries a non-tongue roller and has eight children. If his children exactly meet the expected percentages for tongue rolling and non-tongue rolling, how many of them can roll their tongues?

    a.   0
    b.   1
    c.   2
    d.   4
    e.   8

19. **What is the relationship of Sarah to Debbie?**

    a. twins
    b. homozygous tongue rollers
    c. sisters
    d. heterozygous tongue rollers
    e. not enough information is given

20. **The digestive system of earthworms is much more advanced than the systems of lower order animals. Earthworms were the first animals to have two openings in their digestive system, one for the intake of food and one for elimination of wastes. Food passes from the mouth through the esophagus to the crop. The crop is a sac that is used for temporary storage. Food then passes into the gizzard, which grinds the food and then sends it to the intestines where the food is digested. The nutrients are absorbed and the wastes are then eliminated.**

    Digestion works better on smaller pieces of material. Which organelle is chiefly responsible for preparing the food for digestion in an earthworm?

    a. crop
    b. esophagus
    c. gizzard
    d. stomach
    e. intestine

Stages of a Disease	Definition
Incubation	Initial infestation followed by a quick multiplication of infecting organisms.
Prodrome	A short period of general symptoms such as headache and fever.
Clinical	Characteristic symptoms appear and a diagnosis can be made.
Decline	Symptoms subside, but a relapse is quite possible.
Convalescence	Recovery occurs, however one may still be contagious to others.

21. **Which stage is the best for going to the doctor so that he can prescribe the proper medicine?**

    a. incubation
    b. prodrome
    c. clinical
    d. decline
    e. convalescence

22. **The length of each stage of a disease can depend on a number of factors. First of all, each disease has a typical cycle. The cycle can be altered by the patient's overall health, whether they have been infected by the disease before, if they have been vaccinated for the disease, and finally by medicines prescribed by a doctor.**

People suffering from AIDS often have shorter incubation periods for diseases than most other people. Why would this be the case?

    a. Because they already have a headache and fever, they can skip that phase.
    b. Because their weakened immune system cannot fight the initial infection, so it spreads very quickly.
    c. Because infections piggy-back on the AIDS virus, creating a new virus.
    d. Because new symptoms are hard to diagnose in an AIDS patient.
    e. Because the AIDS virus releases nutrients that help other infections grow.

23. Food poisoning is commonly caused by two different types of bacteria. The first type is salmonella and is usually found in meat. Poultry is especially susceptible to salmonella poisoning. The other type of bacteria that commonly causes food poisoning is staphylococcus. It commonly infects things made with mayonnaise and custard, such as filled baked goods. The incubation period for staphylococcus is two to four hours, while salmonella sets in from six hours up to two days later.

If you had a picnic lunch and went to bed feeling fine, but the next day woke up with severe cramping from food poisoning, which bacteria would you suspect?

    a.   salmonella
    b.   staphylococcus
    c.   It depends on if you ate meat.
    d.   It depends on if you ate something with mayonnaise.
    e.   It depends on if the meat was kept separate from the other food.

24. There are two main formulas for dealing with electricity. The first is voltage equals current times resistance. The second is power equals voltage times current.

If a TV uses 90 watts of power at a voltage of 120 volts, how many amps of current does it draw?

    a. $\frac{3}{4}$ amp
    b.   1 amp
    c. $1\frac{1}{3}$ amps
    d.   90 amps
    e.   10800 amps

25. **A mole is a common term used by chemists. It is equal to the number of atoms needed to weigh in grams a single atom's atomic weight. For instance, hydrogen has an atomic weight equal to one, and one mole of hydrogen weighs one gram.**

Name	Chemical Symbol	Atomic Number	Most Common Atomic Weight
Hydrogen	H	1	1
Helium	He	2	4
Lithium	Li	3	7
Beryllium	Be	4	9
Boron	B	5	11
Carbon	C	6	12
Nitrogen	N	7	14
Oxygen	O	8	16
Fluorine	Fl	9	19
Neon	Ne	10	20
Sodium	Na	11	23
Magnesium	Mg	12	24

How much would one mole of glucose ($C_6H_{12}O_6$) weigh?

    a.   1
    b.   24
    c.   29
    d.   96
    e.   180

26. **75% of the Earth is currently covered with water, but the total area is constantly changing. At the present time the ice caps at both poles are retreating. This is causing the level of the seas to slowly rise. At one time when the ice caps were much larger a land bridge formed between Alaska and Russia.**

If global warming continues as some scientists predict, what will be one of its effects?

    a. A new land bridge will form from Alaska to Russia.
    b. This time, the land bridge will form from Russia to Alaska.
    c. Low-lying areas will be flooded.
    d. Earthquakes will increase.
    e. The plates will move faster because they are warmer.

27. **All the elements are organized into the periodic table. A long time ago, chemists realized elements had properties that repeated over and over. After some careful study, it was discovered the properties of an element were determined by the number of electrons in its outermost shell. One of the most reactive families (a group of elements that have different properties) is the halogens.**

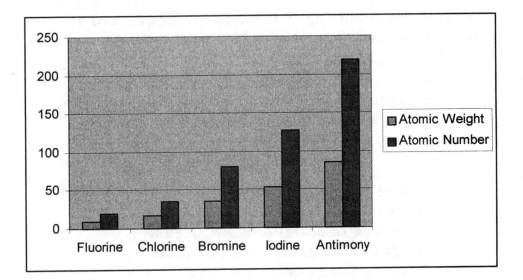

Which halogen has the highest number of neutrons?

   a. Fluorine
   b. Chlorine
   c. Bromine
   d. Iodine
   e. Antimony

28. **The actual origin of cancer is still undetermined but over 80% of the cases can be attributed to avoidable factors, such as ultraviolet or ionizing radiation, exposure to carcinogenic chemicals, and cigarette smoking. Viruses have been shown to cause cancer in animals. Viral particles have been found in some human cancers; however, a direct link between the two has not been proven.**

Which of the following is beyond the control of the average person who is trying to reduce his risk of contracting cancer?

   a. avoiding viruses
   b. reducing ultraviolet radiation exposure
   c. reducing ionizing radiation exposure
   d. reducing exposure to carcinogenic chemical
   e. quitting smoking

WARNING SIGNS FOR CANCER

- A sore that will not heal
- Unusual bleeding or discharge
- A change in bowel or bladder habits
- Obvious change in a wart or mole
- Nagging cough or hoarseness
- A thickening or lump in the breast (or else where)
- Difficulty swallowing or severe indigestion

29. **Which of the warning signs would a physician look for when screening a person for skin cancer?**

   a. Difficulty swallowing
   b. Moles that have changed color
   c. Going to the bathroom frequently in the middle of the night
   d. Hoarseness of the voice
   e. Lumps in the breast

30. **Native Americans used to hunt fish with a bow and arrow. They learned quickly never to aim directly at the fish. Which of the following is the best explanation why they did not?**

   a. The current in the water pushes the arrow away.
   b. The fish will have moved by the time the arrow would get there.
   c. Light is bent by the surface of the water. The fish is not where it appears.
   d. The surface of the water deflects the arrow so it does not travel in a straight path.
   e. The wind blows the arrow before it enters the water.

31. A major way anthropologists date human artifacts is with carbon 14 dating on the organic material found in or with the artifacts. Carbon 14 is a naturally occurring radioactive isotope of carbon. It only makes up a small percentage of all the carbon, but plants incorporate it into glucose along with regular carbon 12. Once it is incorporated into the plant, it can no longer exchange with carbon in the atmosphere. Therefore, after the plant dies, the level of carbon 14 decreases as it undergoes radioactive decay. By looking at the percentage of carbon 14 to carbon 12 it is possible to get an accurate estimation of an object's age.

If carbon 14 has a half-life of 5760 years and an object is 23,040 years old, what fraction of the original carbon 14 remains?

a. $\dfrac{1}{2}$

b. $\dfrac{1}{4}$

c. $\dfrac{1}{8}$

d. $\dfrac{1}{16}$

e. $\dfrac{1}{32}$

32. Scientists discovered a way to refine carbon 14 dating even further. The concentration of Carbon 14 in the atmosphere has varied slightly over time. By taking tree ring samples from bristle cone pine trees (the oldest known living tree), they were able to get accurate carbon 14 levels for the last 5000 years. Fortunately, the environment where bristle cone pines live inhibits decay, so some dead stumps are much older. By matching the growth rings of living trees to dead trees, scientists have been able to go back an additional 5000 years.

Which of the following allows scientists to match rings of living and dead trees?

a. the carbon 14 levels
b. the tree rings in a given area are all affected by such things as a major drought.
c. The sun shone much brighter 5000 years ago.
d. The live trees incorporate the dead trees into their material.
e. The dead trees pass a genetic message to the live trees enabling scientists to decode the older rings.

33. The first thing to do when you find someone unconscious is to call for help. When you are performing CPR you will be working hard and won't be able to yell as loudly. Next try to arouse the victim; do not perform CPR on someone who is breathing or has a heart beat.

A. Airway - Open the airway. Do this by tilting the head back.
B. Breathe – Pinch the nose and give two quick breaths.
C. Circulation – Initiate cardiac compressions.

What is the first thing to do when finding an unconscious victim?

    a.   open the airway
    b.   breathe
    c.   call for help
    d.   begin cardiac compressions
    e.   check for a pulse

Percent of Elements in the Crust

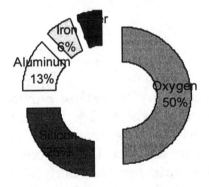

34. Oxygen is about four times more abundant in the crust than which other element?

    a.   Iron
    b.   Aluminum
    c.   Silicon
    d.   Calcium
    e.   Nitrogen

35. According to the "Protoplanet Theory," when gas in the universe combines to form a new sun, eddies (similar to whirlpools) form around it. As the eddies cool, they form into planets.

Why might the inner planets be composed of heavier elements than the outer planets?

    a. The gravitational pull is stronger on planets closer to the sun.
    b. The outer planets had to be lighter so we could call them gas giants.
    c. So that the outer planets could develop rings.
    d. So the outer planets could be bigger.
    e. The asteroid belt keeps the heavy elements from moving outward.

36. **Glasses play an important role in correcting human vision. Look at this diagram to see how using glasses adjusts where light strikes the back of the eye.**

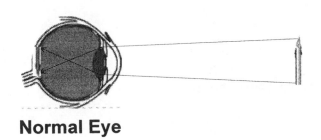

Normal Eye

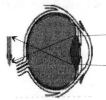

Far-Sighted Eye

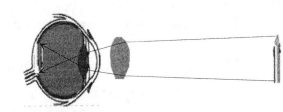

Far-Sighted Eye

How do glasses make it possible for a far-sighted person to see?

    a. They allow people to see candles in the dark.
    b. They keep the candle from getting in your eye.
    c. They correctly focus objects on the back of the eye.
    d. They stretch the eye back to the normal shape.
    e. They shrink the eye to bring the object into focus.

37. Isoleucine is an amino acid essential for growth in infants and nitrogen balance in adults.

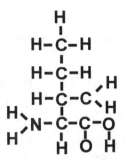

What is the chemical formula for isoleucine?

a. HCNO
b. $C_5N_2O_2H_{12}$
c. C_6NOH_{12}
d. $C_6NO_2H_{11}$
e. $C_6H_{12}NO_2$

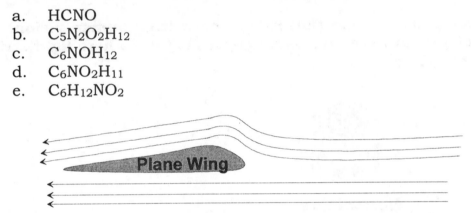

38. The air above the wing has to go farther than the air below the wing. This reduces the pressure of the air above the wing.

Which of the following is true?

a. Low pressure air pushes the plane forward.
b. High pressure air pushes the plane forward.
c. Low pressure air pushes the plane up.
d. High pressure air pushes the plane up.
e. None of the above

39. The Milky Way galaxy is about 80,000 light years in diameter. Our solar system is about ¾ of the way out on one of the two arms that spiral off the central portion of the galaxy. The central cluster has a much higher concentration of stars than the spiral arms.

How far from the Earth is the center of the galaxy?

    a. 80,000 light years
    b. 60,000 light years
    c. 40,000 light years
    d. 30,000 light years
    e. 20,000 light years

40. **Which of the following is a balanced equation?**

    I. $2NaCl + F_2 \rightarrow 2NaF + Cl_2$

    II. $H_2 + O_2 \rightarrow H_2O$

    III. $C + O_2 \rightarrow CO_2$

    a. I only
    b. II only
    c. III only
    d. I and III only
    e. I, II and III

41. **Chemical reactions can be shifted one way or another based on external factors. An exothermic reaction gives off heat, while an endothermic reaction absorbs heat. Other reactions give off gases. If the pressure increases in this type of reaction, it would shift back toward the reactants to relieve pressure.**

A reaction that speeds up when put above a lit Bunsen burner would most likely be what type of reaction?

    a. exothermic
    b. contain hydrogen
    c. pressure sensitive
    d. caloric sensitive
    e. endothermic

42. If plates move, why are earthquakes not happening continuously?

a. They are we just cannot feel them.

b. The plates have an average speed, however most of the time they are stuck. Earthquakes occur when they slip suddenly.

c. Most of the time the slipping does not cause an earthquake because the slippage occurs under water.

d. Earthquakes only occur when the slippage occurs in a north-south direction.

e. Earthquakes can only occur when the temperature is above 32^0 Fahrenheit.

43. One problem with laundry detergent containing phosphates is algae in ponds often uses the phosphates for fertilizer. This causes the algae population to explode. Then the algae die off and all the oxygen in the pond is used up decomposing the dead algae.

Why did the fish in the pond die?
   a. There was no algae to eat.
   b. The same thing that killed the algae killed the fish.
   c. The lack of oxygen in the pond caused them to suffocate.
   d. The algae were poisonous to the fish.
   e. The phosphates were poisonous to the fish.

44. Proteins are the most important chemicals in the cell. One of the most important functions they perform is to act as enzymes. Enzymes are chemicals that speed up reactions. They do this by bringing the reactants close together. This allows the reaction to proceed much easier.

What is the most important way a cell uses proteins?
   a.   For protection
   b.   For food
   c.   For energy
   d.   For waste products
   e.   For enzymes

45. One problem with establishing a permanent base on the moon is getting fresh supplies of water and oxygen. Scientists now suspect that there might be some fairly large deposits of ice at the south pole of the moon. When a colony is founded, the colonies will be able to use the ice for drinking water and to separate the water into hydrogen and oxygen by electrolysis. This will allow the colonists to keep their air fresh.

How will the colonists use ice found on the moon?
   a.   for fresh drinking water
   b.   for construction
   c.   for fuel for rockets
   d.   for fresh air
   e.   both a and d

LANDSLIDE WARNING SIGNS

- Cracks appear in ground, driveways or streets
- Doors or windows stick for the first time
- Telephone poles, fences, or retaining walls start to lean
- New cracks appear in walls, ceiling or foundation
- Walks or stairs begin pulling away from the building
- The ground bulges at the base of a hill
- Underground utility lines break
- Water breaks through the surface in new locations
- Rumbling noise as the landslide approaches

46. **Which of the following is not a warning sign for a coming landslide?**

    a.   five days of continuous rain
    b.   cracks appearing in walls
    c.   the ground bulging at the base of a hill
    d.   telephone poles leaning
    e.   cracks appearing in roadways

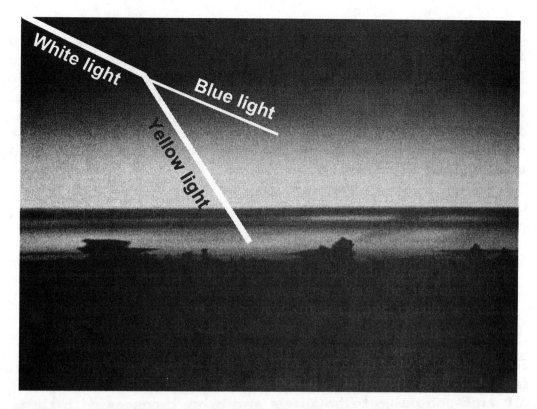

47. The light emitted from the sun is very nearly white. When this light hits the atmosphere, it is refracted and broken into some of its components the same way a prism breaks light into the colors of the rainbow. The blue light gets bent in such a way so that it colors the sky. The rest of the light continues downwards. This is why the sun appears yellow. White light minus blue light equals yellow light.

What color is sunlight in outer space?

    a.   Yellow, just like on earth
    b.   Blue, just like the sky
    c.   Green, a combination of yellow and blue
    d.   White, it hasn't been broken yet
    e.   All the colors of the rainbow

48. Roller coasters are amazing pieces of engineering and physics. Designers calculate every twist and turn and then add large safety margins. The same force that pushes you against the side of a car when you go around a turn keeps the coaster on the track, even when it is upside down. Objects in motion want to keep going in the same direction. When the track turns it must push the car into a new direction. If this force is greater than the force of gravity, the coaster is able to travel upside down.

Which of the following would be another example of this type of force?

   a. someone being unable to jump out of a diving plane
   b. a boat slowly sinking into the sea
   c. a swallow flying in circles to catch bugs
   d. a car going up and over a hill
   e. a train slowing down to go over a bridge

49. Hospitals have unfortunately become a breeding ground for new infectious diseases. This has come about for two reasons. First of all, sick people go to hospitals and bring diseases with them. Some bacteria are able to pass information to each other. If one bacterium is resistant to a type of antibiotic, it can pass that information to other bacteria that have traditionally been treated with that antibiotic. The second reason is many people in hospitals have compromised immune systems. Therefore, an infection a healthy person would normally fight off is able to defeat a compromised immune system.

Which of the following is not true about hospitals and infections?

   a. hospitals try to prevent infections
   b. hospitals try to breed resistant bacteria
   c. hospitals treat people with compromised immune systems
   d. bacteria can transfer information to each other
   e. bacteria can breed quickly

50. What is the best way for hospitals to reduce the number of bacteria that develop a resistance to antibiotics?

   a. allow only healthy people to be admitted
   b. apply fungicide to the floors twice a day
   c. treat resistant infections aggressively to wipe them out before they can transfer any information
   d. put all people with infections together
   e. spray virucide through the air vents daily

Sample Test Answers

1. e. First you have to convert the chemical name to the chemical symbol. Sodium chloride would be written as NaCl. NaCl makes up about 65% of the dissolved solids in seawater.

2. c. When pool balls strike each other, the momentum is conserved. If the first ball stops when they hit, the second goes faster than if they are both moving after the collision.

3. d. Mountain lions eat rabbits and deer. Therefore, they would be classified as carnivores.

4. b. Always make sure you pick the best answer. While both deer and rabbits are shown to eat grass and crops, that is not an answer you can choose. The best answer is autotrophs. Both rabbits and deer are herbivores.

5. b. When the Earth's interior cools, all plate movement will stop because the plates will no longer be floating on a liquid layer.

6. a. Movement is important for most animals. For this reason, it would be impractical for them to limit their movement by developing cell walls.

7. b. The candle would go out because the oxygen is exhausted. Without oxygen, it is impossible for a candle to burn.

8. c. Since rivers are less salty than oceans, they would lower the salinity at the mouth of the river.

9. a. When water reaches 100^0 C it starts to boil. All the energy then goes into converting the water from the liquid state to the gaseous state. The temperature of the pot will not increase until all the water is gone from the pot.

10. d. All three blocks are lifted the same amount, so the work done would be equal.

11. e. The shadow of the Earth always points directly away from the sun. In order for it to fall on the moon, the moon's bright side must be pointed toward the Earth.

12. a. The most abundant compound would be the one with the highest bar on the chart.

13. c. The inorganic molecules are the ones that cannot be made by the cells. They are water and minerals.

14. a. The turtle and the shark exhibit bilateral symmetry.

15. b. The snail does not have any symmetry.

16. a. The solution shown is very close to the low end of the scale. Low numbers are acids; therefore, Solution A is very acidic.

17. b. Both of Tim's parents are non-tongue rollers. To be a non-tongue roller you must have both recessive genes. Therefore, both parents can only pass on recessive genes. This means Tim must be a non-tongue roller.

18. d. First you must determine the genotype of Fred. Since his mother is a non-tongue roller (has two recessive genes), and he is a tongue roller, Fred must have one of each type of gene. If he marries a non-tongue roller, his children will get their characteristics from the gene they get from him. Therefore, half his children should be tongue rollers. The correct answer is four.

19. c. Sarah and Debbie are sisters. They are both on the same line underneath the same set of parents in the chart.

20. c. The passage states the gizzard is where the food is ground. This makes digestion much more efficient.

21. c. You want to go to the doctor when your disease reaches the clinical phase. If you go earlier than that you would just have general symptoms and the doctor would not be able to prescribe medicine for your specific disease.

22. b. AIDS is a disease which attacks the immune system. The immune system is the bodies natural defense system. When infections attack someone with a weakened defense system the infections are able to spread quickly.

23. a. Staphylococcus usually strikes within four hours. Since this attack of food poisoning took longer than four hours to strike, you would suspect salmonella as the culprit.

24. a. Power equals current times voltage. If you want to find the current you must divide the power by the voltage.

$$\frac{90}{120} = \frac{3}{4}$$

25. e. To find the answer to this problem you must find the atomic weight of glucose. Six carbons have an atomic weight of 72. Twelve hydrogens have an atomic weight of 12. Six oxygens have an atomic weight of 96. Adding the three together, we find the atomic weight of glucose to be 180. Therefore, one mole of glucose weighs 180 grams.

26. c. When the Earth's surface gets warmer, the ice caps at the poles tend to melt. When the ice above land melts it will raise the level of the oceans. Ice floating in the water does not affect sea level for the same reason melting ice cubes do not change the level in a glass of water.

27. e. The number of neutrons in an element can be found by subtracting the atomic number from the atomic weight. Looking at the graph, you can see the difference is greatest for antimony.

28. a. Avoiding viruses is the least controllable factor. Many of the viruses which might cause cancer have not even been identified. The other risk factors listed all can be reduced. Ultraviolet radiation can be reduced by wearing sunscreen or simply staying out of the sun. Ionizing radiation can be reduced by avoiding unnecessary x-rays. Avoiding smoking and carcinogenic chemicals will reduce the other two factors

29. b. If you look at the list of warning signs, the ones most related to the skin are festering sores and changes in a mole or freckle.

30. c. The light is bent when it hits the water's surface in the same way that a prism bends light.

31. d. Carbon 14 would go through four half-lives in 23040 years.
$23040 \div 5760 = 4$ $\qquad \frac{1}{2} \times \frac{1}{2} \times \frac{1}{2} \times \frac{1}{2} = \frac{1}{16}$

32. b. When there is a major drought, none of the trees in the area would have a good growing season. Therefore, the rings would be very small. By matching the different poor growing seasons, a scientist is able to match the living and dead trees.

33. c. Calling for help is the first thing to do when you find an unconscious victim.

34. b. Oxygen makes up 50% of the elements in the crust. $50 \div 4 = 12.5$. The element closest to 12.5% is aluminum (13%).

35. a. Since gravitational pull depends on mass and heavier elements have more mass, the heavier elements should be found closer to the sun. This works in the same way a centrifuge can be used to separate blood cells from plasma.

36. c. The problem with a far-sighted eye is objects focus beyond the back of the eye. Glasses adjust where the objects focus so that they can fall on the back of the eye.

37. e. All you need to do for this problem is to add up all the different elements. There are six carbons, twelve hydrogens, two oxygens, and one nitrogen.

38. d. High-pressure air would push up from the bottom of the wing, lifting the plane.

39. d. There are two spiral arms, so the farthest any star would be from the center is 40,000 light years. Since our solar system is only ¾ of the way out, it would be $\frac{3}{4} \times 40,000 = 30000$.

40. d. Equation I has two Na on both sides, two Cl on both sides, and two F on both sides. It is in balance. Equation II has two H on both sides, but two O on one side and only one on the other. It is not in balance. Finally, equation III has one C on both sides and two O on both sides. It is also in balance. Therefore, the correct answer is d-I and III only.

41. e. You need a reaction that occurs faster when it can take energy out of the environment more easily. Therefore, you would want an endothermic reaction.

42. b. When scientists report the plates move at a rate of two inches a year, they are giving the average speed. The plates may be stuck for a number of years and then suddenly slip one foot. These sudden slippages are what causes earthquakes.

43. c. The decomposing algae took all the oxygen out of the water and the fish ended up suffocating.

44. e. The most important way for cells to use proteins is for enzymes. The cell has other sources of food and energy, but only proteins act as enzymes.

45. e. The paragraph states two ways the colonists will use water: for drinking and for fresh air.

46. a. While five days of continuous rain might occur and cause a landslide, it is not always an immediate warning sign before every landslide.

47. d. The paragraph states the sun emits white light. Also, the picture shows white light before it strikes the Earth's atmosphere.

48. a. People can get trapped in a crashing plane because they are unable to overcome the momentum which is trying to keep them going in the same direction.

49. b. While resistant bacteria have developed in hospitals, it was not because the hospitals were trying to breed them. The resistant bacteria came about in spite of the hospitals best efforts to stop them.

50. c. Hospitals must treat resistant bacteria even more aggressively to wipe them out quickly. This way, the resistant bacteria cannot pass their information on to other bacteria.

GLOSSARY

ABSOLUTE ZERO   The temperature at which matter has lost all its thermal energy. [0°K, - 273°C, - 459.7°F].

ACCRETION   The growing together of plant or animal tissues that are normally separate.

ACID   A substance in which H acts as a metal (+ ion). Also, any substance with a pH less than seven.

ADP   Adenosine diphosphate, a compound involved with energy storage and release in the cell.

ADRENAL GLAND   Located on the top of the kidneys. It regulates potassium and sodium in the blood, heartbeat, blood pressure, and blood sugar level.

AEROBIC   Relating to a living organism that requires atmospheric oxygen.

AGAR   A gel prepared from the cell walls of various red algae. It is used in laboratories as a culture medium to grow bacteria.

AIR MASS   A body of air having common characteristics.

ALGAE   Any of numerous groups of eukaryotic (having a nucleus) one-celled or colonial organisms that contain chlorophyll. They usually flourish in aquatic or damp environments. They lack true roots, stems, or leaves.

ALIMENTARY CANAL   The food tube in animals.

ALLELE   One of a pair of genes.

ALLOY   A material composed of two or more metals.

ALTIMETER   An instrument used to measure altitude.

ALVEOLI   The functional unit of the lungs. The tiny bunched air sacs at the end of the bronchioles of the lungs.

AMEBACYTES   A cell that has properties resembling those of an amoeba.

AMMETER   An instrument used to measure the flow of electricity.

AMOEBA   Any of numerous one-celled aquatic or parasitic protozoa of the order Amoebida. They have a jellylike mass of cytoplasm that forms temporary pseudopodia. They use pseudopodia to move and capture food.

AMP   A unit of current equaling one coulomb per second.

AMPERES	See amp.
ANAEROBIC	Relating to organisms unable to survive in atmospheric oxygen.
ANAPHASE	The stage in mitosis or meiosis following metaphase in which the chromosomes move away from each other to opposite ends of the cell.
ANEMOMETER	An instrument that measures wind speed.
ANGSTROM	A linear measurement equal to 1×10^{-8} (0.00000001 cm).
ANHYDROUS	Without water.
ANNELID	Any segmented worm of the phylum Annelida, which includes earthworms and leeches.
ANODE	The positive electrical terminal in an electric cell or battery.
ANTHER	The pollen-bearing part of the stamen.
ANTIBIOTIC	A substance produced by molds, fungi, or bacteria that kills bacteria.
AORTA	The large artery leaving the heart to the body.
APOGEE	The point at which an orbiting satellite is farthest from earth.
AQUIFER	A layer of rock that holds H_2O.
ARACHNID	Any of numerous wingless, carnivorous arthropods of the class Arachnida. It consists of spiders, scorpions, mites, and ticks. They have a two-segmented body with eight appendages and no antennae.
ARCHAEOPTERYX	The first bird.
ARTERIOLES	Any of the smallest branches of an artery.
ARTERY	Large blood vessels carrying blood away from the heart.
ASCUS	The sac in which the spores are formed.
ASEXUAL REPRODUCTION	Reproduction without sperm and eggs.
ASTER	A structure formed in a cell during mitosis, composed of astral rays radiating about the centrosome.
ASTHENOSPHERE	The region below the lithosphere where rock is less rigid than that above and below it.

ASTRONOMICAL UNIT   Distance between the earth and sun—93,000,000 miles.

ATOM   The smallest component of an element having the chemical properties of the element. It consists of a positively charged nucleus of neutrons and one or more electrons in motion around it.

ATOMIC WEIGHT   The relative weight of an atom, compared to the standard $-C^{12}$.

ATP   Adenosine triphosphate, a compound involved with energy use in the cell.

AURORA AUSTRALIS   The aurora of the Southern Hemisphere.

AURORA BOREALIS   The aurora of the Northern Hemisphere.

AXON   A part of the nerve cell that conducts electrical impulses away from the cell body.

BACTERIA   Any of the numerous groups of microscopic one-celled organisms constituting the phylum Schizomycota, of the kingdom Monera. Various species are involved in infectious diseases, nitrogen fixation, fermentation and putrefaction.

BAROMETER   An instrument used to measure air pressure.

BASE   A substance that combines with hydroxyl ions. Also, a substance with a pH above seven.

BIG BANG THEORY   The theory that states the universe began with an explosion of a dense mass of matter. The universe is still expanding from the force of that explosion.

BILE   A bitter, alkaline, yellow or greenish liquid, secreted by the liver, that aids in absorption and digestion, especially of fats.

BINARY STAR   Two stars revolving around each other.

BONE MARROW   The soft fatty vascular tissue in the cavities of bones: a major site of blood cell production.

BOOK LUNGS   The respiratory organ of many arachnids, composed of thin, paper-like layers of tissue.

BOWMAN'S CAPSULE   A membranous, double-walled capsule surrounding a glomerulus of a nephron.

BOYLE'S LAW   The principle stating the pressure of an ideal gas kept at constant temperature varies inversely with the volume of the gas.

BROWNIAN MOVEMENT	The erratic movement of particles due to collisions with atomic or molecular matter.
BTU	British Thermal Unit, a unit of heat energy needed to raise the temperature of one pound H_2O one degree F.
BUDDING	The emergence or branching from the main body of certain simple organisms, as sponges and yeasts. It develops asexually into a new individual.
BUFFER	A substance that will resist any change in the pH of a solution.
BUGS	Insect eggs that hatch into small adult forms, also called nymphs.
CALORIE	A unit of heat energy needed to raise one gram of H_2O one Celsius degree.
CAPILLARY	One of the minute blood vessels between the terminations of the arteries and the beginnings of the veins. This is the site where the exchange of food, gases, and waste occurs.
CATALYST	A substance that controls the rate of a chemical reaction.
CATHODE	Negative electrical terminal in an electric cell or battery.
CELL	The microscopic structure containing nuclear and cytoplasmic material enclosed by a semi-permeable membrane and, in plants, a cell wall; the basic structural unit of all organisms.
CELL MEMBRANE	The semi-permeable membrane enclosing the cytoplasm of a cell.
CENTRAL NERVOUS SYSTEM	The brain and spinal cord.
CENTRIFUGAL REACTION	The reaction of an object to centripetal force.
CENTRIOLE	The small cylindrical organelle in a cell. It is seen near the nucleus in the cytoplasm of most eukaryotic cells. It divides perpendicularly during mitosis.
CENTRIPETAL FORCE	The force that causes a body to conform to movement around a curve.

CENTROMERE	A structure appearing on the chromosome during mitosis or meiosis, where the chromatids are joined in an X shape.
CEPHALOPOD	Any mollusk of the class Cephalopoda, having tentacles attached to the head, including the squid, octopus, and nautilus.
CEREBELLUM	The part of the brain involved with balance and muscle coordination.
CEREBRUM	The part of the brain involved with intelligence.
CHARLES' LAW	The volume of a gas is directly proportional to its temperature (when the pressure remains constant).
CHITIN	A nitrogen containing polysaccharide, related chemically to cellulose. It forms a semi-transparent horny substance. It is a principal constituent of the exoskeleton, or outer covering, of insects, crustaceans, and arachnids.
CHLOROPHYLL	Green substance used in the food manufacturing process in plants.
CHLOROPLASTS	An organelle containing chlorophyll.
CHORDATES	Belonging or pertaining to the phylum Chordata. It comprises the true vertebrates and those animals having a notochord.
CHROMATIDS	Either of two identical chromosomal strands into which a chromosome splits before cell division.
CHROMATIN	The readily stainable substance of a cell nucleus. It consists of DNA, RNA, and various proteins, and forms chromosomes during cell division.
CHROMOSOME	One of a set of threadlike structures, composed of DNA and protein, that forms in the nucleus when the cell begins to divide. It carries the genes that determine an individual's hereditary traits.
CHRONOMETER	A ship's clock which is used to determine longitude.
CIRCUIT	The complete path of an electric current.
CLITORIS	The small erectile organ of the vulva.
CLOT	A semisolid mass, as of coagulated blood.
COELOM	A space forming the body cavity of an animal.

COLD-BLOODED	Animals, such as fishes and reptiles, whose blood temperature ranges from the freezing point upward, in accordance with the temperature of the surrounding environment.
COLLAR CELLS	The individual cells making up the inner cell layer in sponges.
COLLOID	A particle suspended in a medium (most often, liquid).
COLOR	The property of light related to wavelength.
COMPOUND	A chemical composed of two or more parts.
CONDENSATION NUCLEI	The "seeds" upon which H_2O molecules will condense, forming cloud droplets.
CONDUCTION	The transfer of heat between two parts at different temperatures.
CONES	The reproductive structure of certain non-flowering trees and shrubs, as the pine. They consist of hard or papery scales baring naked seeds and arranged in an overlapping whorl around an axis.
CONIFER	Any of a class Pinopsida. They are chiefly evergreen trees and shrubs, as types of the pine and cypress families. Both seeds and pollen are formed in cones.
CONVECTION	The transfer of heat by the circulation or movement of a liquid or gas. Also, the vertical transport of atmospheric properties, especially in an upward direction.
CORIOLIS EFFECT	The deflection of the atmosphere and ocean currents due to the earth's rotation.
COTYLEDON	A thick food leaf formed from the seed of a newly germinated plant.
COVALENT BOND	The bond formed by the sharing of a pair of electrons by two atoms.
CROSS-POLLINATION	The movement of pollen from one plant to another plant.
CRUSTACEAN	Any chiefly aquatic arthropod of the class Crustacea. They typically have the body covered with a hard shell, including lobsters, shrimps, crabs, barnacles, and wood lice.
CRYOGENICS	The science of low temperature phenomena.
CURIE	A unit of radioactivity.

CUTICLE	The outer, non-cellular layer of an arthropod. Also, the very thin waxy film covering the surface of plants, derived from the outer surfaces of the epidermal cells.
CYANOPHYTA	Blue-Green algae.
CYTOPLASM	The cell substance between the cell membrane and the nucleus.
DAUGHTER ELEMENT	An isotope formed by radioactive decay of another isotope.
DDT	A toxic compound, $C_{14}H_9Cl_5$, formerly widely used as an insecticide.
DECIBEL	A unit of sound level intensity.
DECIDUOUS	Relating to woody plants that lose their leaves in winter.
DENDRITE	A part of the nerve that conducts electrical impulse to the cell body.
DENSITY	$D = \dfrac{m}{v}$, mass per unit volume.
DEW POINT	The temperature at which the air is saturated with water vapor.
DIAPHRAGM	Muscular separation between the chest and abdominal cavities.
DIATOMIC MOLECULE	Two atoms in a molecule.
DICOTYLEDON	Two thick food leaves formed from the seed of a newly germinated plant.
DIFFUSION	The spreading out of particles to fill a space.
DIPLOID	Having two copies of each chromosome.
DNA	Deoxyribonucleic acid. A large molecule that controls cell functions. Contains the cell's genetic information.
DOMINANT	Of or pertaining to that allele of a gene pair that masks the effect of the other when both are present in the same cell or organism.
DOPPLER EFFECT	The apparent change in the wavelength caused by movement toward or away from the sound source.
DORSAL	Relating to upper or "back" surface of an animal.
DOUBLE HELIX	The spiral arrangement of the two complementary strands of DNA.

DOUBLE REPLACEMENT REACTION   Two compounds are combined and two products result.

DYNE   A unit of force; 1 dyne = $\dfrac{1\ \text{gram}}{\text{cm}\,/\,\text{sec}^2}$.

E. COLI   Escherichia coli is a common bacterium.

ECHINODERMS   Any marine invertebrate animal of the phylum Echinodermata. It includes starfish and sea urchins, which have radial symmetry and an endoskeleton.

ECOLOGY   The study of the relationship between organisms and their environment.

ECTODERM   Outer layer of cells in an animal body.

EFFERVESCENCE   The rapid escape of gas from a liquid (i.e. soda).

EFFICIENCY   The comparison of work input of a machine to its work output.

EJACULATORY DUCT   A duct through which semen is ejaculated, especially the duct in human males that passes from the seminal vesicle and vas deferens to the urethra.

ELEMENT   One of a class of substances that cannot be separated into simpler substances by chemical means.

ENDOCARDIUM   The membrane that lines the cavities of the heart.

ENDODERM   Inner layer of cells in an animal body.

ENDOPLASMIC RETICULUM   A network of tubular membranes within the cytoplasm of the cell, occurring either with a smooth surface or studded with ribosomes, involved in the transport of materials.

ENDOTHERMIC   The absorption of heat energy during a chemical reaction.

ENERGY LEVEL   The location around the nucleus in an atom where electrons are found.

ENZYME   A protein catalyst which speeds up chemical reactions in cells (suffix is often "ase").

EPICENTER   Point on the surface of the earth that is directly above the focus of an earthquake.

EPIDERMAL   Of the outermost layer of the skin, covering the dermis.

EPIDIDYMIS	An oval structure at the upper surface of each testicle. It consists of tightly convoluted sperm ducts.
EQUILIBRIUM	A situation in which two opposite processes occur at the same rate (water 32°F ice).
EQUINOX	The days when the sun's direct rays strike perpendicularly at the equator (March 21 and September 21).
ERG	Unit of work; 1 erg = 1 dyne × 1 cm.
EROSION	Movement of sediments by wind, water, and ice.
ERYTHROCYTE	A red blood cell.
ESOPHAGUS	A muscular tube for the passage of food from the pharynx to the stomach; gullet.
ESTROGEN	Any of several major female sex hormones produced primarily by ovarian follicles. It induces estrus, produces secondary female sex characteristics, and prepares the uterus for the reception of a fertilized egg.
EUGLENA	Any freshwater protozoan of the genus Euglena. They have a reddish eyespot and a single flagellum.
EUTROPHICATION	Aging of ponds or lakes as plants and sediments fill them in.
EVAPORATION	To change from a liquid or solid state to into vapor.
EVOLUTION	The gradual change of organisms with respect to acquired traits.
EXCURRENT	Giving passage outward.
EXOSKELETON	An external covering, as the shell of a crustacean.
EXOTHERMIC	The release of heat energy during a chemical reaction.
EXTRUSIVE ROCKS	Igneous rocks that form on or near the surface.
FALLOPIAN TUBE	Either of a pair of long slender ducts in the female abdomen that transport ova from the ovary to the uterus. Also, they transport sperm cells from the uterus to the released ova.
FAULT	A break in rocks along which movement has occurred.
FERTILIZATION	The joining together of sperm and egg.
FETUS	The young of an animal in the womb or egg, especially in the later stages of development. In human beings this is after the end of the second month of gestation.
FIBRINOGEN	A protein present in blood and aids in clotting.

FIRN	Snow or ice pellets that become part of a glacier.
FISSION	The breakup of a heavy nucleus into two smaller nuclei.
FLAGELLUM	A long lash-like appendage used for locomotion in protozoa, sperm cells, etc.
FOCUS	The point in the Earth's crust where an earthquake originates.
FORCE	That action which produces or prevents changes in motion.
FOSSIL	Remains or evidence of life as it was in the past.
FOVEA	A part of the eye which has many cones.
FREQUENCY	Number of cycles per unit time.
FRICTION	A force that opposes motion.
FROND	A large, finely divided leaf, found in ferns and certain palms. Also, a leaf-like expansion not differentiated into stem and foliage, as in lichens.
FRONT	The line along which two air masses collide.
FRUIT	It develops from the ovary in a plant.
FULCRUM	A pivot point.
FUSION	The combining of two lightweight nuclei into one heavier nucleus.
GALVANIZE	To coat iron or steel with Zinc.
GAMETOPHYTE	A stage in reproduction in which gametes are formed.
GASTRIC JUICES	The digestive fluid, containing pepsin and other enzymes, secreted by the glands of the stomach.
GASTROPOD	Any of numerous mollusks of the class Gastropoda. Snails, whelks, and slugs have a single coiled shell or no shell at all. They move by means of a wide muscular foot.
GASTRULA	The early stage of development when germ layers appear.
GEAR	A disk, wheel, or section of a shaft, having cut teeth that mesh with teeth in another part to transmit or receive force and motion.
GEIGER COUNTER	An instrument that detects radioactivity.
GENE	A portion of the DNA molecule that produces a trait in the organism.

GENETIC CODE	The chemical sequence in DNA that produces a trait.
GENOTYPE	The genetic makeup of an organism or group of organisms with reference to a single trait or set of traits.
GEOSYNCLINE	The downfold of the earth's crust which is being filled with sediments.
GEOTROPISM	Response of plants to gravity.
GEYSER	Steam and hot water that erupt from cracks in the Earth's surface.
GILL	An organ used to remove oxygen from water.
GIZZARD	The thick-walled, muscular lower stomach of many birds and reptiles that grinds partially digested food.
GLACIER	A moving body of ice.
GLOMERULUS	Any compact cluster of nerves or capillaries, especially a cluster of capillaries in the nephron of the kidney that acts as a filter of the blood.
GLUCOSE	A simple sugar.
GOLGI BODY	One of the layers of flattened sacs in a Golgi apparatus.
GONADS	Male and female reproductive organs.
GREENHOUSE EFFECT	The heating of the atmosphere resulting from gases trapping the heat radiating from the Earth's surface.
HAPLOID	An organism or cell having only one complete set of chromosomes.
HARD WATER	Water containing dissolved minerals.
HEART	A muscular organ that receives blood from the veins and pumps it through the arteries.
HEAT OF FORMATION	The amount of heat needed or released when a compound is formed.
HERBACEOUS	Relating to the stem of a plant that lives only one season.
HERBIVORES	Plant-eating animals.
HEREDITY	The passing of traits from parents to offspring.
HERMAPHRODITE	An organism in which reproductive organs of both sexes are present.
HETEROZYGOUS	Having dissimilar alleles for hereditary characteristics.

HOMOZYGOUS	Having a pair of identical alleles for a given hereditary character.
HORIZON	A line of vision where the surface meets the sky.
HORMONE	Any of various internally secreted compounds formed in endocrine glands. They affect the functions of specific organs or tissues when transported to them by the body fluids.
HORSEPOWER	A unit of power equal to 550 ft.-lb./sec.
HOST	An organism used by a parasite to supply it with food.
HUMUS	Decayed organic matter.
HYDRA	Any freshwater polyp of the family Hydridae, having a cylindrical body with a ring of tentacles surrounding the mouth.
HYDRATION	The attachment of water molecules to a particle of matter.
HYDROCHLORIC ACID	It is a very strong acid.
HYDROLOGIC CYCLE	The natural cycle through which water passes. It goes from the atmosphere as water vapor, precipitates to earth, and returns to the atmosphere through evaporation.
HYGROMETER	An instrument that determines relative humidity.
HYPHAE	One of the threadlike elements of the mycelium in a fungus.
HYPOTHESIS	A possible explanation that is proved or disproved by an experiment.
IGNEOUS ROCKS	Rocks that form from lava or magma.
IMPERFECT FUNGI	Fungi which only have an asexual reproductive system.
INCLINED PLANE	One of the simple machines used to reduce the amount of force needed to perform work.
INCURRENT	Carrying or relating to an inward current.
INORGANIC	Materials which are *not* hydrocarbons.
INSOLATION	Incoming solar radiation.
INSULATOR	A poor conductor of heat or electricity.
INSULIN	A hormone produced by the pancreas to regulate the oxidation of sugar in cells.

INTERNAL RESPIRATION	The exchange of gases between the cells and the blood.
INTERNATIONAL DATE LINE	The 180° east or west longitude line (bent for convenience) that separates dates east and west (crossing over the dateline east to west—gain a day; west to east—lose a day).
INTERPHASE	The period of the cell cycle during which the nucleus is not undergoing division.
INTRUSIVE ROCKS	Igneous rocks that form deep within the Earth's crust.
INVERTEBRATE	Animals lacking a backbone.
IONIC BOND	The electrostatic bond between two ions formed through the transfer of one or more electrons.
IRIS	Colored portion of the eye.
ISO	Prefix meaning "the same."
ISOBAR	A line connecting points of equal air pressure on a weather map.
ISOTOPE	One of two or more forms of a chemical element having the same number of protons, yet have different numbers of neutrons.
JOULE	Unit of work equaling 1 nt. × 1 m.
KIDNEY	One of a pair of organs in the rear of the upper abdominal cavity of vertebrates that filter waste from the blood, excrete uric acid or urea, and maintain water and electrolyte balance.
KINETIC ENERGY	The energy of motion.
KNOT	One nautical mile/hour.
LACTASE	An enzyme used to breakdown lactose.
LATITUDE	Degrees north or south of the equator.
LAVA	Molten rock that reaches the earth's surface.
LAW OF INDEPENDENT ASSORTMENT	Mendel's conclusion that one trait may be inherited independently from another trait.
LAW OF SEGREGATION	Mendel's conclusion that genes are separated during the formation of the sex cells (meiosis) and that the genes recombine during fertilization.

LEUKOCYTE	A white blood cell.
LEVER	A bar used as a simple machine. It increases the distance through which a force acts.
LICHEN	They are composed of a fungus in symbiotic union with an alga. Most commonly they form crusty patches on rocks and trees.
LIGHT REACTION	The first stage of photosynthesis in which energy is produced and used to split water molecules.
LIPASE	An enzyme used to change fat into fatty acids and glycerin.
LITHOSPHERE	The crust of the Earth.
LOCAL NOON	The time at which the sun crosses the observer's meridian (longitude).
LONGITUDE	Degrees east or west of the Prime Meridian.
LOOP OF HENLE	The part of a kidney tubule that loops from the cortex into the medulla of the kidney.
LUNAR MONTH	Twenty-nine and a half days.
LUNG	An organ used by air-breathing animals.
LUSTER	Shine of a mineral surface (metallic, earthy, pearly, glassy, etc.).
LYSOME	A cell organelle containing enzymes that break down proteins and other large molecules into smaller constituents. They disintegrate the cell itself after its death.
MAGMA	Liquid rock that is located deep within the Earth's crust.
MAGNITUDE	The *apparent* brightness of a star.
MALARIA	Any of a group of intermittent or remittent diseases characterized by attacks of chills, fever, and sweating. They are caused by a parasitic protozoan transferred to the human bloodstream by an anopheles mosquito.
MAMMALS	Any warm-blooded vertebrate of the class Mammalia. They are characterized by a covering of hair on some or most of the body, a four-chambered heart, and nourishment of the newborn with milk from maternal mammary glands.

MAMMARY GLAND	Organs of female mammals that occur in pairs on the chest or ventral surface and contain milk-producing lobes with ducts that empty into a nipple.
MANOMETER	An instrument used to measure gas pressure.
MANTLE	An outgrowth of the body wall in mollusks and brachiopods. It lines the inner surface of the shell valves and secretes a shell-forming substance.
MARE	Latin word meaning "sea."
MARROW	The soft interior part of bones.
MARSUPIALS	An animal of the order Marsupialia. They are mammals having no placenta and bear immature young that complete their development in a pouch on the mother's abdomen.
MASS	The measure of the amount of matter that an object has.
MEDULLA OBLONGATA	The large part of the spinal cord next to the brain.
MEDUSA	The free-swimming body form in the life cycle of a jellyfish or other coelenterate, usually dome-shaped with tentacles.
MEIOSIS	Sex cell formation. The reduction of chromosomes from a diploid number to a haploid number.
MENDEL, GREGOR	(1822-84) An Austrian botanist who is often called "The Father of Genetics."
MENSTRUATION	The periodic discharge of blood and mucosal tissue from the uterus, occurring approximately monthly from puberty until menopause in non-pregnant women and females of other primate species.
MERIDIAN	An imaginary line from pole to pole; longitude line.
MESODERM	Middle layer of cells in an animal body.
MESOGLEA	The non-cellular, gelatinous material between the inner and outer body walls.
MESOZOIC ERA	The age of the reptiles.
MESSENGER RNA	M-RNA. RNA molecules that are fashioned from the DNA in the nucleus of the cell. They travel from the nucleus to the cytoplasm to the site where information will be moved to transfer RNA's.

METAL
Elements that readily donate electrons in chemical reactions; have luster, are good conductors, and are malleable.

METAMORPHIC ROCKS
Rocks formed from changes in sedimentary, igneous, or metamorphic rocks, caused by extreme heat and pressure (short of melting the rock) or contact with igneous magma.

METAMORPHOSIS
Change in the body form of an animal as it goes through its life cycle.

METAPHASE
The stage in mitosis or meiosis in which the duplicated chromosomes line up along the equatorial plate.

MILKY WAY
The spiral galaxy containing our solar system, seen as a luminous band stretching across the night sky and composed of approximately a trillion stars.

MITOCHONDRIA
An organelle in the cell cytoplasm that has its own DNA. They are inherited solely from the maternal line. The powerhouse of the cell produces enzymes essential for energy metabolism.

MITOSIS
Identical cell reproduction.

MIXTURE
Two or more substances combined physically. May be separated by physical means.

MOHO
Short for Mohorovicic discontinuity, the zone separating the crust from the mantle in the earth.

MOLLUSK
An invertebrate of the phylum Mollusca. They can have a calcareous shell of one or more pieces that wholly or partly encloses the soft, unsegmented body.

MOLTING
The shedding of the outer covering of skin, skeleton (arthropods), or feathers (birds).

MOMENTUM
(Mass) × (velocity) of an object equals its momentum.

MONOCOTYLEDON
A newly germinated plant with only one food leaf.

MONOTREME
Any egg-laying mammal of the order Monotremata. The duckbill and the echidnas of Australia and New Guinea are the only known varieties.

MORAINE
Rocks deposited by glacial movement.

MOTOR NEURON
A nerve cell that carries electrical impulses from the central nervous system to a muscle or organ.

MUCUS A solution of water, electrolytes, and white blood cells. It is secreted by mucous membranes and serves to protect and lubricate the internal surfaces of the body.

MUTANT A cell containing non-inherited traits. These traits appear from a mutation caused by radiation or some other source.

MYELIN SHEATH A discontinuous wrapping of myelin around certain nerve axons. They increase the nerve impulses.

MYOCARDIUM The muscular substance of the heart.

MYRIAPODA Arthropods having elongated segmented bodies with numerous paired, jointed legs. It consists of centipedes and millipedes.

NATURAL SELECTION Those traits that allow organisms to compete successfully to survive and produce offspring.

NAUTICAL MILE 1.062 land miles.

NEBULA A cloud of interstellar gas and dust.

NEMATOCYST The stinging cells in jellyfish and other coelenterates.

NEMATODE Any unsegmented worm of the phylum Nematoda. They have an elongated, cylindrical body and are often parasitic on animals and plants; a roundworm.

NEPHRON The filtering and excretory unit of the kidney, consisting of the glomerulus and convoluted tubule.

NEUTRAL Having no net electrical charge.

NEUTRON An elementary particle found in most atomic nuclei. They have no charge and a mass just slightly greater than that of a proton.

NODE OF RANVIER Gaps in the myelin sheath occurring at regular intervals along a nerve axon.

NON-METAL Elements that accept electrons easily in chemical reactions; poor conductors.

NOTOCHORD A long, flexible structure in chordates and vertebrate embryos. In vertebrates it develops into the spinal column.

NOVA A star that is expanding rapidly and giving off a tremendous amount of light energy.

NUCLEAR MEMBRANE A double membrane surrounding the nucleus within a cell.

NUCLEOLUS	A small, rounded body within the cell nucleus. It is responsible for ribosome manufacture.
NUCLEUS	A specialized, spherical mass of protoplasm found in most cells. It directs their growth, metabolism, and reproduction. The nucleus contains most of the genetic material.
NYMPH	The young of an insect that undergoes incomplete metamorphosis.
OBLATE SPHEROID	A sphere flattened at the poles and bulging at the equator (shape of the Earth).
OHM	The unit of electrical resistance.
OLFACTORY	Having to do with smell.
OOZE	Fine sediments found on the sea floor.
ORBIT	The path of a body that revolves about another body (electrons around a nucleus, Earth around the sun).
ORGAN	A group of tissues that perform certain functions.
ORGANIC	Relating to carbon compounds; hydrocarbons.
OSMOSIS	The movement of water through a membrane from an area of high concentration to an area of lower concentration.
OSSIFICATION	The change from cartilage to bone.
OVA	The plural of ovum. The female reproductive cell (egg) developed in the ovary.
OVARY	The female reproductive organ.
OVIPAROUS	Egg-laying animals.
OXIDATION	The process by which oxygen combines with other substances.
PALTSADW	The cells responsible for photosynthesis in plants.
PANCREAS	A large compound gland, situated near the stomach, that secretes digestive enzymes into the intestine. It also produces glucagon and insulin to regulate blood sugar levels.
PARAMECIUM	A freshwater protozoan of the genus Paramecium. They have an oval body with an oral groove and are covered with cilia for locomotion.
PARASITE	A living organism that is dependent upon another living organism for its food.

PARATHYROID GLAND Small paired glands in vertebrates lying near the thyroid gland. They secrete parathyroid hormone.

PASCAL'S PRINCIPLE Pressure in a confined fluid acts equally in all directions.

PELECYPOD It means "two shells;" a bi-valve mollusk.

PENUMBRA Partial shadow caused by an eclipse.

PERICARDIUM The membranous sac enclosing the heart.

PERIGEE The point at which an orbiting object is closest to the point it is moving around.

PERMEABLE ROCK Rock that has enough interconnected pore spaces so as to allow water to pass through it.

pH The H^+ ion count present in a solution. Range 1-14; 1-6.9 is acid ($H^+ > OH^-$), 7.0 is neutral ($H^+ = OH^-$), 7.1 - 14 is base ($OH^- > H^+$).

PHENOTYPE The traits an organism exhibits (what it looks like).

PHLOEM The tubes of a plant which bring the food from the leaves down to the roots.

PHOTOSYNTHESIS The production of glucose from carbon dioxide and water. It uses sunlight as the source of energy.

PINEAL GLAND A small gland in the posterior forebrain. It produces melatonin and is involved in biorhythms and gonadal development.

PISTIL The seed-bearing organ of a flower. It consists of the ovary, style, and stigma.

PITUITARY GLAND A small gland attached to the base of the brain. It is often called the master gland and affects all hormonal functions of the body.

PLACENTAL Animals that have a placenta.

PLANULA The free-swimming larva of coelenterates.

PLASMA The fluid part of the blood or lymph, not the cellular components.

PLASTIC A material that can be shaped while soft.

PLATE A block of crustal rock (basalt-granite) which sits on top of the asthenosphere.

PLATELETS The smallest cells in the blood. They are disk-shaped and contain no hemoglobin. They are essential to the clotting of blood.

PLATYHELMINTH	Any of various unsegmented worms of the phylum Platyhelminthes. They have soft, flattened bodies and include tapeworms and planarian.
POLYMER	A compound formed by joining two organic compounds in a repeating chain.
POTENTIAL ENERGY	Energy due to the position of an object.
PRECIPITATE	A product that settles out of solution during a chemical reaction.
PRIME MERIDIAN	0° longitude passes through Greenwich, England.
PROGESTERONE	A female hormone formed in the corpus luteum of the ovary. It causes the lining of the uterus to prepare for a fertilized egg.
PROMINENCE	A large solar flare that erupts from the sun's surface.
PROPHASE	The first stage of mitosis or meiosis in cell division. During this stage, the nuclear envelope breaks and strands of chromatin form into chromosomes.
PROSTATE GLAND	A partly muscular gland that surrounds the urethra in males. It secretes an alkaline fluid that makes up part of the semen.
PROTEIN	Foods made of carbon, oxygen, hydrogen, and nitrogen.
PROTHALLUS	The gametophyte phase of ferns and related plants.
PROTON	A positively charged elementary particle found in all atomic nuclei.
PROTOZOAN	Any of various one-celled protist organisms that usually obtain nourishment by ingesting food particles rather than by photosynthesis.
PSEUDOPOD	A temporary protrusion of the cytoplasm. It is usually used for locomotion or to capture food.
PSYCHROMETER	An instrument that is used to determine relative humidity.
PTYALIN	An enzyme that changes starch to sugar.
PULLEY	A wheel for supporting, guiding or transmitting force to or from a moving rope or cable that rides in a groove in its edge.
PULSAR	A star that varies regularly in brightness.
QUANTUM	An elemental unit of energy.

QUASAR	Quasi-stellar radio source.
RADIATION	The process in which energy is emitted as particles or waves.
RADIOACTIVE DECAY	In a given amount of time, half of a radioactive substance will decay into daughter elements.
RADIOSONDE	Weather instrumentation package sent up in a balloon.
RADULA	A tongue-like band in the mouth of most gastropods.
RAIN GAUGE	A collecting device that measures rainfall.
RECEPTOR	A group of cells that receive a stimulus.
RECESSIVE	Pertaining to the allele of a gene pair whose effect is masked by the second allele when both are present in the same cell or organism.
REDUCTION	The removal of oxygen from a compound.
REGENERATION	The formation of new body parts.
RELATIVE HUMIDITY	The percentage of water vapor in the air at a particular temperature.
RESISTANCE	The opposition to the flow of current. This causes electrical energy to be changed into heat.
RESPIRATION	The exchange of gases in cells. Also, it is the act of breathing.
RHIZOME	A root-like underground stem commonly found in ferns. They usually produce roots and send up shoots.
RIBOSOME	The organelle in cells where proteins are produced.
RNA	Ribonucleic acid. Carries directions to various parts of the cell as directed by the DNA.
SALIVA	A watery fluid secreted into the mouth by the salivary glands. It moistens the mouth and starts the digestion of starches.
SALT	A compound formed by (+) and (-) ions.
SAPROPHYTE	An organism that obtains its food from non-living organic matter.
SATELLITE	Any body revolving around another body.
SCROTUM	The pouch of skin that contains the testes.
SECOND LAW OF MOTION	Force = Mass x Acceleration, or F = MA

SEDIMENTARY ROCKS	Rocks formed from clastic, organic, or chemical precipitates or evaporates.
SEED	A plant embryo.
SEISMOLOGY	The study of earthquakes.
SELF-POLLINATION	The process in which pollen fertilizes ova of the *same* plant.
SEMEN	A whitish fluid produced in the male reproductive organs, containing sperm.
SEMINAL VESICLE	A sac-like gland located on each side of the bladder in males. They add nutrient fluid to semen during ejaculation.
SENSORY NEURON	The part of the nerve that carries electrical impulses.
SERIES	An arrangement of an electrical circuit in which the components are connected end-to-end, so that the same current flows through each component.
SETAE	Stiff hair or bristles.
SEXTANT	An instrument used to determine the altitude of a celestial body.
SEXUAL REPRODUCTION	Reproduction involving sperm and egg.
SILICA	Quartzsand, flint, and agate; used chiefly in the manufacture of glass, water glass, ceramics, and abrasives.
SINGLE REPLACEMENT REACTION	Chemically, when a compound and an element are combined to form a new product.
SIPHON TUBE	A projecting part of certain mollusks, through which liquid enters of leaves the body.
SLIME MOLDS	Any of various fungus-like organisms belonging to the phylum Myxomycota of the kingdom Protista. They have an amoeboid phase and a streaming phase.
SMOG	Mixture of air; pollutants (i.e. smoke) and fog.
SOLAR WIND	Particles that travel away from the sun's surface.
SOLSTICE	The northernmost or southernmost advance of the vertical rays of the sun (summer, Northern Hemisphere, 23 1/2 °N; winter, northern hemisphere, 23 1/2 °S).

SOLUTE	The dissolved substance in a solution.
SOLUTION	A mixture of two or more substances.
SOLVENT	The dissolving medium.
SORI	Clusters of sporangia on the back of the fronds of ferns.
SPECIES	A group of living organisms that have the same characteristics and reproduce with each other.
SPECIFIC GRAVITY	Ratio of the weight of a substance to the weight of an equal volume of water.
SPECTROSCOPE	An instrument used to split white light into the colors of the rainbow (Remember the acronym, ROY G. BIV— red, orange, yellow, green, blue, indigo, and violet).
SPERM	A male reproductive cell.
SPICULE	Small, hard, calcareous (or siliceous) skeletal elements of various marine and freshwater invertebrates.
SPIRACLE	One of the external orifices of the respiratory system in certain invertebrates.
SPLEEN	A highly vascular ductless organ located near the end of the stomach. It stores red blood cells, destroys old red blood cells, and makes new white blood cells.
SPORE	The asexual reproductive body of a fungus or non-flowering plant.
SPOROPHYTE	The generation of some plants that produces asexual spores.
STALACTITE	A mass of calcite hanging from the ceiling of a cave.
STALAGMITE	A mass of calcite on a cave floor building up under a stalactite.
STAMEN	The pollen-bearing organ of a flower, consisting of the filament and the anther.
STEM	The part of a plant, whether above or below ground, which ordinarily grows in an opposite direction to the root.
STOMATA	Minute openings in leaves, stems, etc., through which gases are exchanged.
STP	Standard temperature and pressure; 760 mm of Hg (1atm) and 0°C.

STRIATED	Striped or streaked. Also, the type of muscle under conscious control.
STYGMA	The sticky top of a pistil in flowering plants.
STYLE	A narrow, cylindrical extension of the pistil.
SUBLIME	Changing from the solid state to the gaseous state without passing through the liquid state.
SUBSCRIPT	Written below.
SUCRASE	An enzyme that breaks down sucrose.
SUPERPOSITION	Younger rocks lie on top of older rocks.
SYMBIOSIS	Two organisms that live together in order to survive. Both organisms must benefit from the relationship.
SYNAPSE	The space between nerve endings.
SYNTHETIC	Man-made.
SYSTEM	A group of organs that perform a specific function.
TALUS	Rock material at the foot of a cliff.
TELOPHASE	The final stage of mitosis or meiosis in cell division, during which the two sets of chromosomes reach opposite poles. The nuclei form around the chromosomes as the cell divides into two new cells.
TENTACLE	Any of various slender, flexible appendages in animals. They serve as organs of touch or prehension.
TERRESTRIALS	Land-living organisms.
TESTES	The male reproductive organ.
THERMOCLINE	A layer of water separating the warm and cold water regions of a large body of water.
THIRD LAW OF MOTION	The law stating that for every action there is an equal and opposite reaction.
THROMBOCYTE	One of the minute cells that aid clotting in the blood of those vertebrates which do not have blood platelets.
THYMUS	A ductless gland at the base of the neck. It is formed mostly of lymphatic tissue and aids in the production of T cells for the immune system.
THYROID GLAND	A two-lobed endocrine gland at the base of the neck. It secretes two hormones that regulate the rates of metabolism, growth, and development.
TIME MERIDIAN	Meridian located at the center of a time zone.

TRACHEA	The air tube in terrestrial arthropods. Also, part of the air passageway in air-breathing vertebrates.
TRANSFER RNA	T-RNA. Fashioned from M-RNA blueprint, T-RNA attracts and connects amino acids to form proteins.
TRANSFORMER	A device that changes the amount of voltage in a circuit.
TRANSPIRATION	The passage of water through a plant from the roots through the vascular system to the atmosphere.
TRENCH	A deep depression on the ocean floor.
TROPOSPHERE	The lowest layer of the atmosphere, varying in height from six to twelve miles, within which nearly all clouds and weather conditions occur.
TSUNAMI	A tidal wave generated by earthquakes.
TUBE FEET	Small, tubes on the lower body surface of most echinoderms. They are used for locomotion and grasping.
UMBRA	The dark shadow caused by an eclipse.
UNIFORMITARIANISM	Understanding the present is the key to understanding the past.
URASIL	A nucleic base which found in RNA. It forms base pairs with adenine.
UREA	A nitrogen waste product formed from used proteins.
URETHRA	A duct that conveys urine from the bladder to the exterior. In most male mammals, it also conveys semen.
URINE	The waste matter excreted by the kidneys.
VACUOLE	A membrane-bound cavity within a cell, often containing a watery liquid or secretion.
VAGINA	The passage leading from the uterus to the vulva in female mammals.
VAN ALLEN BELT	Either of two atmospheric regions of high-energy charged particles, one at an altitude of about 2,000 miles and the other from about 9,000 to 12,000 miles.
VASCULAR BUNDLES	Strands of xylem and phloem found in the higher plants.
VAS DEFERENS	A duct that transports sperm from the epididymis to the penis.

VEIN	One of the system of branching vessels or tubes conveying blood from various parts of the body to the heart.
VENA CAVA	Either of two large veins discharging blood into the right atrium of the heart.
VENOUS	Of or pertaining to a vein or veins.
VENTIFACT	A wind abraded rock.
VENTRAL	Relating to "front" or lower surface of an animal.
VENTRICLE	Any of various hollow organs or parts in an animal body. Also, either of two lower chambers of the heart that receive blood from the atria and, in turn, force it into the arteries.
VENULE	A small vein. Also, one of the branches of a vein in the wings of an insect.
VERTEBRATE	Having a segmented backbone.
VESTIGIAL ORGANS	Non-functioning organs.
VIABLE	Capable of surviving outside the uterus.
VILLI	Projections on the wall of the intestine which absorb nutrients.
VIRUS	A microscopic non-living infectious agent. It replicates only within the cells of living. It is composed of an RNA or DNA core and protein coat.
VISCOSITY	Internal friction of a fluid.
VOLATILE	Easily vaporized.
VOLT	Unit of electrical potential difference between two points in an electric field.
VOLTAGE	Electromotive force or potential difference expressed in volts.
VULCANIZATION	Application of heat to rubber products.
VULVA	The external female genitalia.
WARM-BLOODED	Animals, as mammals and birds, having a body temperature that is relatively constant and independent of the environment.
WATERSHED	An area drained by one major stream.

WATER VASCULAR SYSTEM The rhythmic movement of water through the body that allows animals (such as starfish) to move through water.

WATT Unit of power; 1 amp × 1 volt = 1 W.

WEATHERING The process of breaking up rocks by chemical and/or physical means.

WHEEL AND AXLE A simple machine which overcomes force due to friction.

WOMB The uterus.

WORK (Force)(distance) = work

X-CHROMOSOME The sex determining chromosome. Two x chromosomes make a female. One x and one y chromosome make a male.

XYLEM The tubes in plants conducts water and nutrients upward from the roots to the leaves.

ZENITH The point in the sky directly above an observer's head.

ZYGOSPORES A cell formed by fusion of two similar gametes, as in certain algae and fungi.

ZYGOTE The cell produced by the union of two gametes (sex cells), before it undergoes division.

ZYMASE Enzyme in yeast cells that acts on sugar to produce alcohol and carbon dioxide.